AF851255

Progress in Mathematics
Vol. 50

Edited by
J. Coates and
S. Helgason

Swiss Seminars

A subseries of Progress in Mathematics
Adviser: A. Borel

This subseries is devoted to Notes emanating from seminars held in Switzerland, mostly in Bern, with the participation of researchers in several Swiss academic institutions. They are primarily devoted to expositions of topics of much current interest and they aim at facilitating access to them.

Birkhäuser
Boston · Basel · Stuttgart

A. Borel *et al.*

Intersection Cohomology

1984

Birkhäuser
Boston · Basel · Stuttgart

22. JUL. 1985

Author:

A. Borel
Mathematik and School of Mathematics
ETH-Zentrum The Institute for Advanced Study
CH–8092 Zürich Princeton, NJ 08540
(Switzerland) (USA)

Library of Congress Cataloging in Publication Data
Main entry under title:

Intersection cohomology.
(Progress in mathematics ; v. 50. Swiss seminars)
"Notes of a seminar on intersection homology, which met weekly during the spring 1983 at the University of Bern, Switzerland"--Forew.
Bibliography: p.
1. Homology theory. 2. Intersection theory. 3. Piecewise linear topology. 4. Sheaves, Theory of. I. Borel Armand. II. Universität Bern. III. Series: Progress in mathematics (Boston, Mass) ; v. 50. IV. Series: Progress in mathematics. Swiss seminars.
QA612.3.I58 1984 514'.2 84-20347
ISBN 0-8176-3274-3

CIP-Kurztitelaufnahme der Deutschen Bibliothek

Intersection cohomology / A. Borel et al. - Boston ; Basel ; Stuttgart : Birkhäuser, 1984.
(Progress in mathematics ; Vol. 50 : Swiss Seminars)
ISBN 3-7643-3274-3 (Basel ...)
ISBN 0-8176-3274-3 (Bosten ...)
NE: Borel, Armand [Mitverf.]; GT

515.142
BOR

AMM 222

All rights reserved.
No part of this publication may be reproduced, stored in a retrieval system, or transmitted, in any form or by any means, electronic, mechanical, photocopying, recording or otherwise, without prior permission of the copyright owner.

© Birkhäuser Boston, Inc., 1984
ISBN 0-8176-3274-3
ISBN 3-7643-3274-3
Printed in Switzerland
9 8 7 6 5 4 3 2 1

FOREWORD

This volume contains the Notes of a seminar on Intersection Homology which met weekly during the Spring 1983 at the University of Bern, Switzerland. Its main purpose was to give an introduction to the piecewise linear and sheaf theoretic aspects of the theory (M. Goresky and R. MacPherson, Topology 19(1980)135-162, Inv. Math. 72(1983) 17-130) and to some of its applications, for an audience assumed to have some familiarity with algebraic topology and sheaf theory.

These Notes can be divided roughly into three parts. The first one (I to IV) is chiefly devoted to the piecewise linear version of the theory: In I, A. Haefliger describes intersection homology in the piecewise linear context; II, by N. Habegger, prepares the transition to the sheaf theoretic point of view and III, by M. Goresky and R. MacPherson, provides an example of computation of intersection homology. The spaces on which intersection homology is defined are assumed to admit topological stratifications with strong local triviality properties (cf I or V). Chapter IV, by N. A'Campo, gives some indications on how the existence of such stratifications is proved on complex analytic spaces.

The primary goal of V is to describe intersection homology, or rather cohomology, in the framework of sheaf theory and to prove its main basic properties, following the second paper quoted above. Familiarity with standard sheaf theory, as in Godement's book, is assumed. However, this paper makes use of considerably more material on constructibility, derived categories, Verdier duality and biduality. A second goal of V is then to supply an essentially self-contained account of what is required. This material is gradually introduced, according to the needs of the discussion of intersection cohomology. As a consequence, some auxiliary results are proved first only in special cases. To compensate for that, a section on various identities in derived categories of sheaves has been added.

VI, by P.P. Grivel, is devoted to some basic properties of direct and inverse images functors and complements on some points §§ 1,6,7 of V .

The third part (VII to IX) contains a description by M. Goresky of some work of P. Siegel of cobordism, the statement of a Lefschetz fixed point theorem in intersection cohomology (M. Goresky and R.MacPherson) and finally a discussion of open problems and a comprehensive bibliography on intersection homology, also by M. Goresky and R. MacPherson.

To complete the picture, I should at least mention an important item left out of these Notes, the so-called "decomposition theorem", pertaining to middle intersection cohomology of projective varieties, whose proof, at this time, makes use of characteristic p methods which were outside the scope of this seminar (see ref. [A] of IX). In fact, R. MacPherson gave a lecture on the decomposition theorem and on some of its uses to make effective computations, but no written version was included in the Notes, since it was felt it would be too much of a duplication with existing or forthcoming literature (see e.g. ref.[20] in IX).

A preliminary version of these Notes was circulated beforehand, with the hope of eliciting comments and corrections. This was particularly successful with Ludger Kaup and two of his colleagues at Konstanz university, namely G. Barthel and K.-H. Fieseler, who read it meticulously and pointed out a great number of misprints and oversights, for which the authors of the various chapters are very grateful.

In drafting my own part, I was also very fortunate to benefit from the help of N. Spaltenstein. He checked various versions of the text with extreme care and suggested many changes, whether corrections, simplifications, alternate arguments or additions, which have considerably improved the text. I am very grateful to him for his assistance, which goes way beyond the few changes which have been specifically attributed to him in the text, and has much speeded up the preparation of the final text.

Finally, I would like to thank the editors of the Progress in Mathematics PM and Birkhäuser-Boston for their willingness to view this volume as the first of a subseries emanating from activities in Switzerland, although it must be said that, at this point, it cannot be guaranteed that it will have (m)any successor(s).

A. Borel

Zurich, July 1984

TABLE OF CONTENTS

Intersection Cohomology
Progress in Mathematics, Vol. 50
© 1984 Birkhäuser Boston, Inc.

I : INTRODUCTION TO PIECEWISE LINEAR INTERSECTION HOMOLOGY

by A. Haefliger

1. *Pseudomanifolds and stratifications*

1.1 *Topological stratified pseudomanifolds.* The definition is by induction on the dimension n. A *stratified pseudomanifold X of dimension n* is a topological space X with a filtration

$$X = X_n \supset X_{n-2} \supset X_{n-3} \supset \ldots \supset X_1 \supset X_0 \supset X_{-1} = \emptyset$$

by closed subspaces such that

(i) $S_{n-k} = X_{n-k} - X_{n-k-1}$ is a topological manifold of dimension n-k (if S_{n-k} is non empty).

(ii) $X - X_{n-2}$ is dense in X.

(iii) local normal triviality : for each point $x \in S_{n-k}$, there is a compact stratified pseudomanifold L of dimension k-1

$$L = L_{k-1} \supset L_{k-3} \supset \ldots \supset L_0 \supset L_{-1} = \emptyset$$

and a homeomorphism h of an open nbhd U of x (called a distinguished nbhd of x) on the product $B \times \overset{\circ}{c}L$, where B is a ball nbhd of x in S_{n-k} and $\overset{\circ}{c}L$ is the open cone $L \times [0,\infty[/(x,0) \sim (x',0)$ over L. Moreover h preserves the stratifications, namely h maps homeomorphically $U \cap X_{n-\ell}$ on $B \times \overset{\circ}{c}\, L_{k-\ell-1}$ (by definition, the cone on the empty set is just a point).

[library stamp]

X_{n-2} is often called the *singular locus* Σ of the stratified pseudomanifold X.

1.2 *Example.* Let X be the algebraic variety in C^3 defined by the equation

$$y^2 z = x^2$$

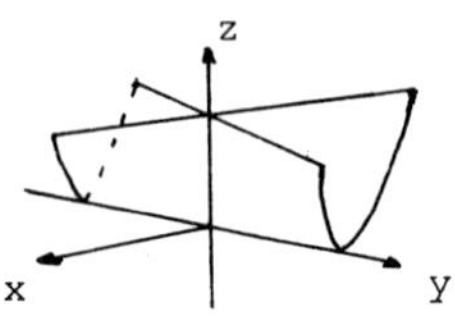

The singular locus $\Sigma = X_2$ is the complex line x=y=0. To get the normal local triviality, we need to include the origin in X_0. Note that $X-X_2$ is dense in X in the complex picture.

1.3 *Pl-pseudomanifolds.* Recall that a *pl-space X* (i.e. piecewise-linear) is a topological space with a class of locally finite simplicial triangulations of X : two admissible triangulations T and T' should have a common linear subdivision and any linear subdivision of T should be admissible.

Any open set U of X has an induced pl-structure : for any admissible triangulations T of X and T_U of U, then there is a linear subdivision of T_U such that each simplex is contained linearly in a simplex of T. A closed pl-subspace of X is a subspace which is a subcomplex of a suitable admissible triangulation of X.

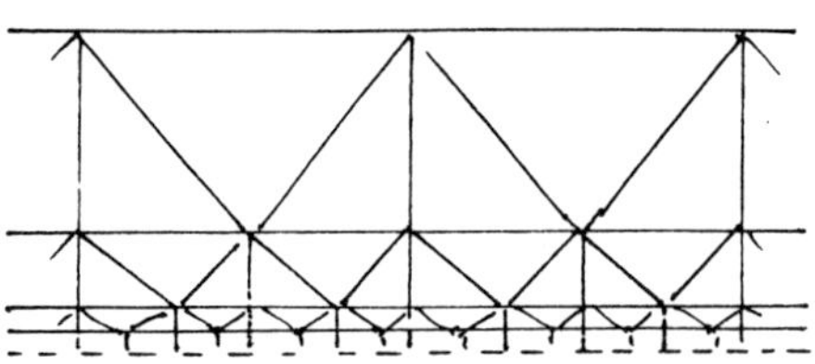

picture showing a triangulation of the open half-plane

A *pl-pseudomanifold X* of dimension n is a pl-space X of dimension n containing a closed pl-subspace Σ of codimension ≥ 2 such that $X-\Sigma$ is a pl-manifold of dimension n dense in X.

Equivalently, for an (admissible) triangulation of X, then X is the union of the closed n-simplices and each (n-1)-simplex is face of exactly two n-simplices.

A *stratified pl-pseudomanifold X* of dimension n is a pl-pseudomanifold with a stratification $X \supset X_{n-2} \supset X_{n-3} \supset \ldots$ by closed pl-subspaces such that $S_{n-k} = X_{n-k} - X_{n-k-1}$ is empty or a pl-manifold of dimension n-k and such that the local normal triviality holds in the pl-category (namely L is a pl-stratified pseudomanifold and h is a pl-homeomorphism).

1.4 *PROPOSITION : Any pl-pseudomanifold admits a pl-stratification.*

Proof. Let T be a triangulation of X such that Σ is a subcomplex of T, and define X_k as the union of the closed k-simplices of dimension $\le$ k contained in Σ. Hence $X_{n-k} - X_{n-k-1}$ is the union of the interior of the (n-k)-simplices contained in Σ.

To check the local normal triviality, consider the first barycentric subdivision T' of T. For a (n-k)-simplex σ of T contained in Σ, let $D'(\sigma)$ (resp. $L'(\sigma)$) be the union of those simplices of T' whose vertices are the barycenters $\hat{\tau}$ of the simplices τ of T containing σ (resp. containing σ and different from σ).

$D'(\sigma)$ will be called the dual complex of σ and $L'(\sigma)$ the link of σ in T'. Note that $D'(\sigma)$ is the join $\hat{\sigma} * L'(\sigma)$ of the barycenter $\hat{\sigma}$ of σ with $L'(\sigma)$, so $D'(\sigma)$ is the cone over $L'(\sigma)$ and the interior $\overset{\circ}{D}(\sigma)$ of $D'(\sigma)$ is the open cone on $L'(\sigma)$. Also the union $St'(\sigma)$ of the closed simplices of T' containing the barycenter of σ is the join $\partial\sigma * D'(\sigma)$.

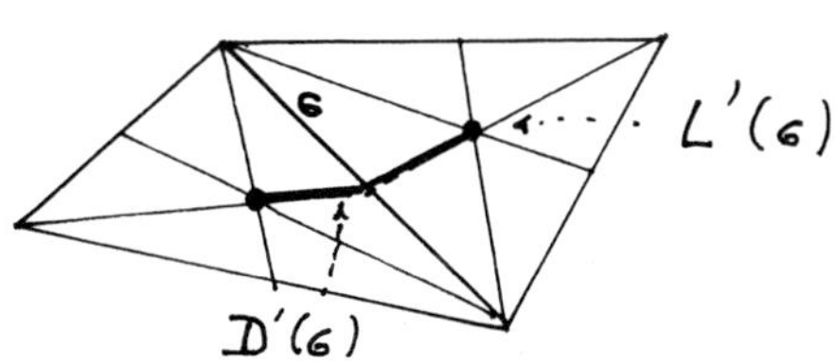

We consider on $L'(\sigma)$ the stratification induced by the stratification of X. The interior of $St'(\sigma)$ is pl-homeomorphic to (interior of σ) $\times \overset{\circ}{D}(\sigma)$ by a pl-homeomorphism compatible with the filtrations.

1.5 *Basic example.* X is a complex analytic space whose irreductible components are all of dimension n over $\mathbb{C}$.

By a theorem of Lojasiewicz [7] there is a semi-analytic triangulation of X such that the singular locus of X is a subcomplex (more generally such that a given locally finite family of complex analytic subspaces of X are subcomplexes). For such a triangulation, X is a pl-pseudomanifold of dimension 2n and it can be stratified as above.

Of course such a stratification is very artificial. Better topological stratifications with even dimension strata can be obtained (cf. [9] p.220 and the exposé of A'Campo, IV). There exist such a pl-stratification on X such that the local normal triviality holds in the pl-category.

Anyway later on it will be proved that the intersection homology is a topological invariant independent of the particular choice of the stratification and of the pl-structure.

1.6 *Normal pseudomanifolds and normalisation.* A pseudomanifold X is *normal* if each point x has a fundamental system of neighbourhoods U whose regular part $U-\Sigma$ is connected.

If X is stratified and if $U \simeq B \times \overset{\circ}{c}L$ is a distinguished nbhd of x, then U is normal iff L is a normal pseudomanifold and is connected.

For any stratified pseudomanifold X, there is a normal pseudomanifold $\tilde{X}$ with a projection $\pi : \tilde{X} \to X$ uniquely characterized by the property that the points of $\pi^{-1}(x)$ are in bijection with the connected components of the regular part of a distinguished nbhd U of x. The stratified pseudomanifold $\tilde{X}$ (where $\tilde{X}_{n-k} = \pi^{-1}(X_{n-k})$) together with the projection $\tilde{\pi} : \tilde{X} \to X$ is called the *normalisation of X.*

If X is a complex analytic space, then the normalisation of X as an analytic space is topologically equivalent to the normalisation in the above sense for the pseudomanifold X.

The normalisation can be constructed by induction on the dimension. If $U = B \times \overset{\circ}{c}L$ is a distinguished nbhd and if the normalisation $\tilde{L}$ of L has been constructed, then the normalisation $\tilde{U}$ of U is the disjoint union of the $B \times \overset{\circ}{c}\tilde{L}_i$, where the $\tilde{L}_i$ are the connected components of $\tilde{L}$.

As an example, the normalisation of the cone over a pinched torus is the cone over a 2-sphere, so is a 3-ball.

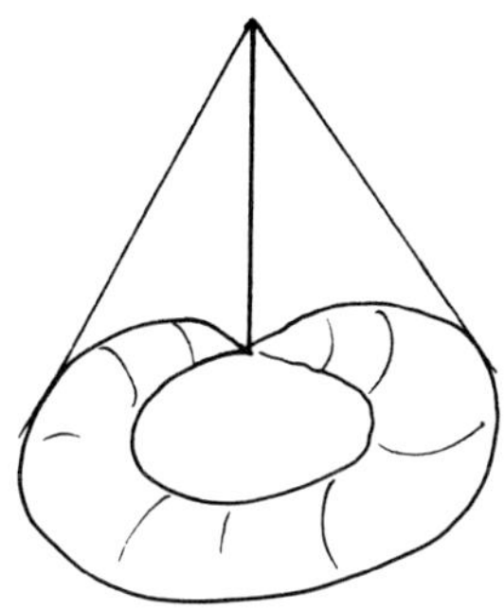

2. *Geometric chains on a pl-space.*

2.1 Let X be a (locally finite) pl-space and R be an abelian group. For a triangulation T of X, we denote by $C_\cdot^T(X;R)$ the complex of (possibly infinite) simplicial chains of T. An element ξ of $C_i^T(X;R)$ is a function associating to each oriented i-simplex σ of T an element $\xi(\sigma)$ of R (and to the simplex σ with the opposite orientation the element $-\xi(\sigma)$). The support $|\xi|$ of ξ is the union of the closed i-simplices σ of T such that $\xi(\sigma) \neq 0$. This is a closed pl-subspace of X.

Let T' be a subdivision of T. There is a natural morphism of complexes $C_\cdot^T(X;R) \to C_\cdot^{T'}(X,R)$ associating to the i-chain ξ the simplicial chain ξ' defined as follows : for an oriented i-simplex σ' of T', then $\xi'(\sigma') = 0$ if σ' is not contained in a i-simplex of T and $\xi'(\sigma') = \xi(\sigma)$ if σ' is contained in the i-simplex σ of T with the coherent orientation. In general for two triangulations T and T' of X we say that $\xi \in C_i^T(X;R)$ and $\xi' \in C_i^{T'}(X;R)$ define the same *geometric chain* if their image in $C_i^{T''}(X;R)$, where T" is a common linear subdivision of T and T' are the same. Roughly speaking, to define a geometric chain ξ, one uses a particular triangulation and then one forgets it.

This equivalence relation preserves the support, so that the *support* $|\xi|$ *of a geometric chain* ξ is a well defined pl-subspace of X. If R is the group of mod 2 integers, then a geometric chain is just a closed pl-subspace, of pure dimension, namely its support.

2.2 *Definition.* The complex of geometric chains of X with coefficients in R will be denoted by $C_{\cdot}(X;R)$ and its homology by $H_{\cdot}(X;R)$ (and called the homology with closed supports of X or the Borel-Moore homology of X).

The subcomplex of geometric chains with compact support will be denoted by $C^c_{\cdot}(X;R)$ and its homology by $H^c_{\cdot}(X;R)$ (and called the homology with compact support of X).

The classical proof that the simplicial homology of X is isomorphic to its singular homology shows that $H^c_{\cdot}(X;R)$ is isomorphic to the homology of the complex of locally finite singular chains (called homology of the second kind in the Cartan Seminar, chapter 5) and also to the Borel-Moore homology of X without restriction on the supports.

As we don't use the standard notation, we emphazise once more that

$H_{\cdot}(X;R)$ denotes the homology with closed supports of X

$H^c_{\cdot}(X;R)$ denotes the usual homology of X.

(Of course if X is compact, these two homologies are the same.)

For instance

$$H_i(\mathbb{R}^n;R) = \begin{cases} R & i=n \\ 0 & i\neq n \end{cases}$$

$$H^c_i(\mathbb{R}^n;R) = \begin{cases} R & i=0 \\ 0 & i\neq 0 \end{cases}$$

If R is a field, then the universal coefficient theorem gives

$$H^i(X;R) = \mathrm{Hom}(H^c_i(X;\mathbb{Z}),R)$$

$$H_i(X;R) = \mathrm{Hom}(H^i_c(X;\mathbb{Z}),R)$$

where $H^{\cdot}$ denotes the singular cohomology.

Any open set U of X has an induced pl-structure (cf.1.3). Let ξ be a geometric chain of dimension i defined using a triangulation T of X; let T_U be a triangulation of U such that each k-simplex of T_U is contained in a simplex of T. The restriction $\xi|U$ of ξ to U is the geometric chain represented by the simplicial chain of T_U which associates to an oriented i-simplex σ of T_U the element $\xi(\tilde{\sigma})$, when $\tilde{\sigma}$ is a i-simplex of T containing σ (with the coherent orientation), and 0 otherwise. In this way, the inclusion $j : U \to X$ induces an homomorphism

2.2.1 $$j^* : H_i(X;R) \to H_i(U;R)$$

called the restriction homomorphism.

Similarly we have an inclusion of the complex of geometric chains with compact supports in U in $C^c_.(X)$ inducing a homomorphism

2.2.2 $$j_* : H^c_i(U;R) \to H^c_i(X;R)$$

For a closed pl-subspace Y of X one can define the complex of relative geometric chains $C.(X,Y;R)$ as $\varinjlim_T C^T_.(X;R)/C^{T|Y}_.(Y;R)$, where the limit is taken over those triangulations T of X for which Y is a subcomplex.

The homology of this limit will be denoted by $H.(X,Y;R)$.

The restriction map $C.(X;R) \to C.(X-Y;R)$ induced by the inclusion of X-Y in X gives a map $C.(X,Y;R) \to C.(X-Y;R)$.

2.2.3 *Fact :* This map induces an isomorphism $H_i(X,Y;R) \simeq H_i(X-Y;R)$ (cf. Borel-Moore [1] . An elementary proof can be given by induction on the simplices of a triangulation T).

2.3 *The fundamental geometric cycle of an oriented pl-pseudomanifold X.* An orientation of a pl-pseudomanifold X of dimension n is an orientation of its regular part. It is given by a coherent orientation of all the n-simplices of a triangulation T of X. In other words, the chain associating $1 \in \mathbb{Z}$ to an n-simplex with the chosen orientation is an n-cycle the corresponding geometric cycle will be denoted by X and called the fundamental cycle of the oriented pseudomanifold X. Its support is the whole of X.

More generally if R is a ring (with unit 1), an R-orientation of X is a geometric cycle [X] with coefficients in R with support X and associating to each oriented n-simplex ± 1.

Its restriction to an open set U of X gives an R-orientation of U.

2.4 *The Poincaré duality map.* Let T be a triangulation of X and let $C_T^{\cdot}(X,R)$ be the complex of simplicial cochains of T with coefficients in R. Let T' be the first barycentric subdivision of T.

An R-orientation [X] of the pseudomanifold X induces a natural morphism of complexes

$$C_T^{n-\cdot}(X:R) \xrightarrow{\cap[X]} C_{\cdot}^{T'}(X,R).$$

It maps the cochain associating $1 \in R$ to the oriented (n-k)-simplex σ and 0 to the others on the k-chain of $C_k^{T'}(X;R)$ with support $D'(\sigma)$ (cf.1.4) and multiplicity one for a suitable orientation ; its boundary is the fundamental cycle of the pseudomanifold $L'(\sigma)$ for a suitable R-orientation. See Mc Crory [8] , for an explicit formula.

This map induces homomorphisms

$$H^{n-k}(X;R) \xrightarrow{\cap[X]} H_k(X;R)$$

$$H_c^{n-k}(X;R) \xrightarrow{\cap[X]} H_k^c(X,R)$$

(more generally it preserves a family of supports).

For a n-manifold, this map induces the Poincaré duality isomorphism.

3. *Definition of intersection homology*

3.1 *Perversity.* A *perversity* p is a function associating to each integer k, $2 \le k \le n$, an integer p(k) such that

$$p(2) = 0 \quad \text{and} \quad p(k) \le p(k+1) \le p(k)+1.$$

The graph of p looks like stairs with steps of height 0 or 1 :

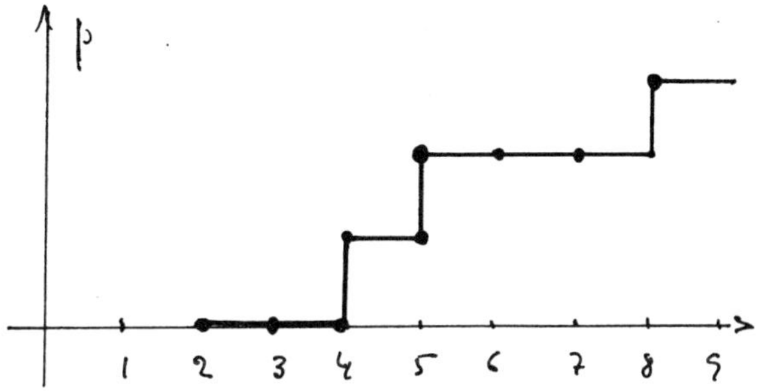

The o-perversity is the function o(k) = 0 and the top perversity t is the function t(k) = k-2.

The middle perversities n or m are the functions taking the values 0,1,1,2,2,... or 0,0,1,1,2,...

Two perversities are complementary if p(k)+q(k) = k-2. For instance o and t are complementary as well as m and n.

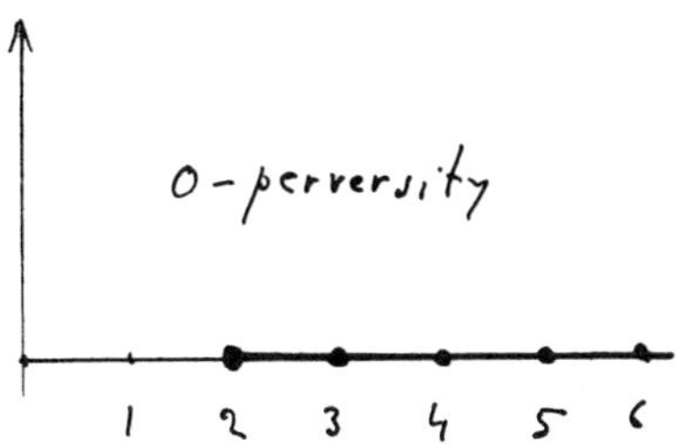

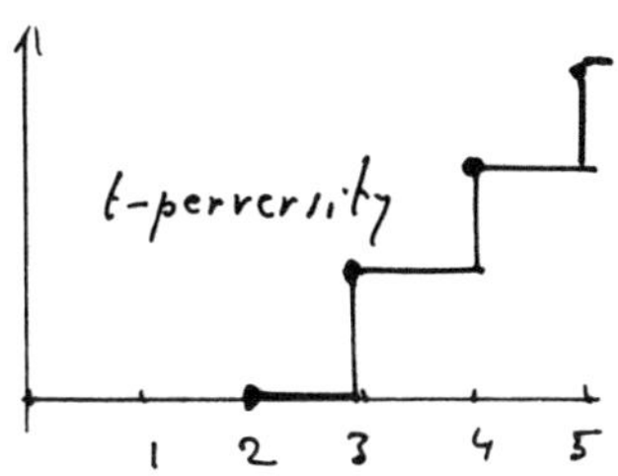

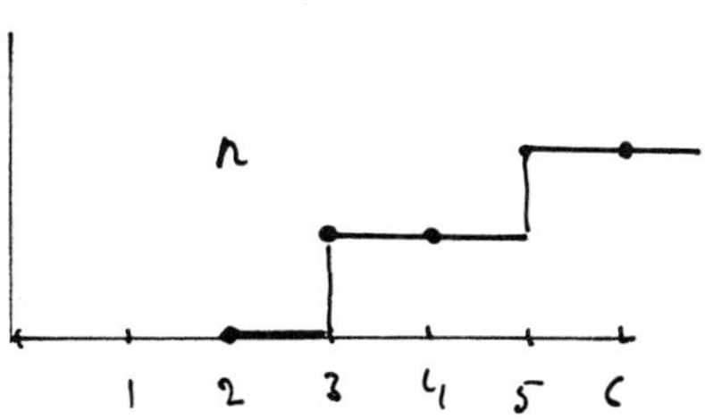

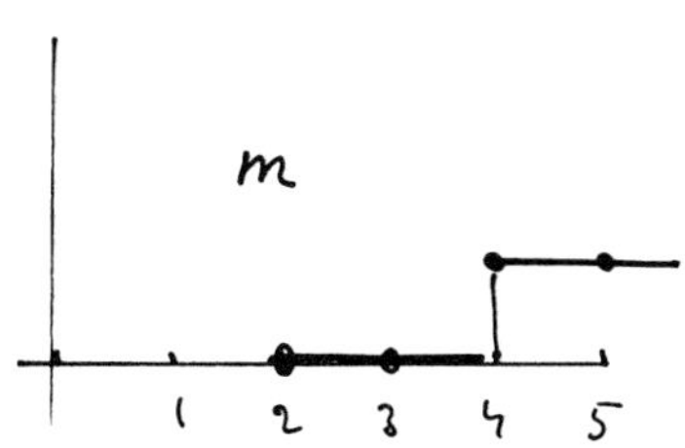

3.2 *Definition of intersection homology.* Let X be a pl-pseudomanifold of dimension n with a given stratification

$$X = X_n \supset X_{n-2} \supset \ldots \supset X_0 \supset \emptyset.$$

The group $I_pC_i(X;R)$ *of intersection chains of dimension i and perversity p is the subgroup of geometric chains* $\xi \in C_i(X;R)$ *such that, for each subspace* X_{n-k},

$$\dim|\xi| \cap X_{n-k} \le i-k+p(k)$$

$$\dim|\partial\xi| \cap X_{n-k} \le i-1-k+p(k)$$

Because of the condition imposed on $\partial\xi$, $I_pC_\cdot(X;R)$ *is a subcomplex of* $C_\cdot(X;R)$ *whose homology will be denoted by* $I_pH_i(X;R)$ (or $I_pH^{n-i}(X;\tau\otimes R)$ in Borel's notation, where τ is the orientation sheaf of the regular part of X).

We make a few obvious remarks to clarify the definition. If X were a manifold, a chain ξ of dimension i in general position with respect to the stratification would satisfy $\dim|\xi| \cap X_{n-k} \le i-k$. So the perversity p(k) is the extra amount of this intersection dimension which is allowed. Note that an intersection chain ξ is never contained in the singular locus $\Sigma = X_{n-2}$ since the codimension in $|\xi|$ of the intersection of $|\xi|$ with Σ is at least 2. In particular any 0 or 1-dimensional intersection chain has its support contained in $X-\Sigma$. Any n-cycle in $C_n(X;R)$ is in $I_pC_n(X;R)$. Two intersection chains in $I_pC_i(X;R)$ are equal iff their restriction to $X-\Sigma$ are equal.

It follows easily that if $\pi : \tilde{X} \to X$ is the normalisation of X, then π induces a bijection of $I_pC_\cdot(X;R)$ on $I_pC_\cdot(X;R)$, so that the intersection homology of $\tilde{X}$ and X are the same.

If p and p' are two perversities such that $p(k) = p'(k)$ for all k such that $X_{n-k}-X_{n-k-1} \ne \emptyset$, then $I_pC_\cdot(X;R) = I_{p'}C_\cdot(X;R)$. In particular if all the strata are of even codimension, then the two middle perversities n and m give the same intersection chains.

One has a natural inclusion $I_pC_i(X;\mathbb{Z})\otimes R\subset I_pC_i(X;R)$ which is not surjective in general (because of the condition on $\partial\xi$), so there is no universal coefficient theorem in general.

For an open set U of X with the induced stratification, the inclusion $j : U \to X$ induces by restriction (resp. inclusion) maps

$$I_pC.(X;R) \to I_pC.(U;R)$$
$$I_pC^c_.(U;R) \to I_pC^c_.(X;R)$$

where $I_pC^c_.$ denotes the subcomplex of intersection chains with compact support. Passing to homology, we get homomorphisms functorial with respect to inclusion

3.2.1 $$j^* : I_pH.(X;R) \to I_pH.(U;R)$$

3.2.2 $$j_* : I_pH^c_.(U;R) \to I_pH^c_.(X;R)$$

where $I_pH^c_.(X;R)$ denotes the homology of the complex $I_pC^c_.(X;R)$.

In particular for the injection j of $X-\Sigma$ in X, we have that

$$j_* : H^c_o(X-\Sigma;R) \to I_pH^c_o(X;R)$$

is an isomorphism and is surjective in dimension one.

For $p\leq p'$ (i.e. $p(k)\leq p'(k)$ for all k), we have an inclusion $I_pC.(X;R) \to I_{p'}C.(X;R)$ inducing a homomorphism

3.2.3 $$I_pH.(X;R) \to I_{p'}H.(X;R) \quad p\leq p'$$

If X is an R-orientation of X (cf.2.3), then, for any triangulation T of X such that the X_{n-k}'s are subcomplexes, the map

$$\cap[X] : C^{n-i}_T(X R) \to C_i(X R)$$

described in 2.4 factorizes through I_pC_i for any p (since $[D'\sigma]\in I_oC_i$) and gives homomorphisms

3.2.4
$$\cap[X] : H^{n-i}(X;R) \to I_pH_i(X;R)$$
$$\cap[X] : H_c^{n-i}(X;R) \to I_pH_i^c(X;R)$$

4. *List of the main properties of intersection homology*

X is assumed to be a stratified pl-pseudomanifold of dimension n.

4.1 *Relation with cohomology and homology.* If X is a normal pseudomanifold, then the natural map

$$I_tH.(X;R) \to H.(X;R)$$

induced by the inclusion $I_tC.(X;R) \to C.(X;R)$ is an isomorphism.

This is also valid for homology with compact support.

If X is R-orientable and if X in an R-orientation, then the natural map

$$\cap[X] : H^{n-i}(X;R) \to I_oH_i(X;R)$$

is an isomorphism, as well as

$$\cap[X] : H_c^{n-i}(X;R) \to I_oH_i^c(X;R)$$

4.2 *Intersection pairing.* Assume that p,q and r are perversities such that

$$p(k)+q(k) \le r(k) \quad \text{for all } k$$

and that X is R-oriented.

Then there are bilinear intersection pairings

$$I_pH_i(X;R) \times I_qH_j(X;R) \overset{\cap}{\to} I_rH_{i+j-n}(X;R)$$
$$I_pH_i^c(X;R) \times I_qH_j^c(X;R) \overset{\cap}{\to} I_rH_{i+j-n}^c(X;R)$$
$$I_pH_i^c(X;R) \times I_qH_j(X;R) \overset{\cap}{\to} I_rH_{i+j-n}^c(X;R)$$

If j is the inclusion of the open set U in X, then the first is compatible with the restriction j^* to U and the second with j_*. For the third one, for $\alpha \in I_pH_i^c(U;R)$ and $\beta \in I_qH_j(X;R)$, one has

$$j_*\alpha \cap \beta = j_*(\alpha \cap j^*\beta)$$

When restricted to the regular part they give the usual intersection product corresponding to the cup product under Poincaré duality.

These intersection products are associative and commutative in the graded sense, namely $\alpha \cap \beta = (-1)^{(n-i)(n-j)}\beta \cap \alpha$.

Representative geometric cycles ξ and η for $\alpha \in I_pH_i(X;R)$ and $\beta \in I_qH_j(X;R)$ can be chosen such that $\dim(|\xi| \cap |\eta|) \leq i+j-n = \ell$ and $\dim(|\xi| \cap |\eta| \; X_{n-k}) \leq \ell-k+r(k)$; then a representative for $\alpha \cap \beta$ has its support contained in $|\xi| \cap |\eta|$.

4.3 *Poincaré duality.* Let p be a perversity and q be the complementary perversity (so $p(k)+q(k) = k-2$).

Assume that R is a *field* and that X is R-oriented.

Then the intersection pairing

$$I_pH_i(X;R) \times I_qH^c_{n-i}(X;R) \xrightarrow{\cap} I_tH^c_o(X;R)$$

composed with the augmentation $I_tH^c_o(X;R) = H^c_o(X-\Sigma) \xrightarrow{\varepsilon} R$ gives a pairing

$$I_pH_i(X;R) \times I_qH^c_{n-i}(X;R) \to R$$

which is non degenerate, in the sense that it induces an isomorphism

$$I_pH_i(X;R) \xrightarrow{\cong} \mathrm{Hom}(I_qH^c_{n-i}(X;R),R).$$

<u>4.4</u> *Independence of the stratification and topological invariance.* $I_pH.(X,R)$ is independent of the stratification of the pl-pseudomanifold X. Moreover $I_pH.(X;R)$ is finitely generated if X is compact.

All of these properties are essentially proved by Goresky and Mac Pherson in their first paper [3] .In their second paper [4] , using sheaf theoretical techniques, they also define intersection homology for topologically stratified pseudomanifolds and prove the independance with respect to the stratification. So the groups I_pH are topological invariants of X.

5. *Example : pseudomanifolds with isolated singularities*

<u>5.1</u> Let X be a pl-pseudomanifold of dimension n whose singular set $\Sigma = X_o$ is of dimension 0.

For a given perversity p, only the integer p(n) is relevant, where

$$0 \leq p(n) \leq n-2$$

Let ϕ be the family of closed sets in X which are contained in $X-X_o$ (so if X is compact, ϕ is the family of compact sets in $X-X_o$). We denote by $C^{\phi}_{\cdot}(X-X_o)$ the complex of geometric chains in $X-X_o$ whose support belongs to ϕ and by $H.(X-X_o)$ its homology.

A ring R of coefficients is understood throughout.

A geometric chain ξ of dimension i will satisfy the perversity condition iff

$$\dim|\xi|\cap X_o \leq i-n+p(n)$$
$$\dim|\partial\xi|\cap X_o \leq i-1-n+p(n)$$

So

$$I_pC_i(X) = \begin{cases} C_i^{\phi}(X-X_o) & i<n-p(n) \\ C_i(X)\cap\partial^{-1}C^{\phi}_{i-1}(X-X_o) & i=n-p(n) \\ C_i(X) & i>n-p(n) \end{cases}$$

hence

5.1.1

$$I_pH_i(X) = \begin{cases} H_i^{\phi}(X-X_o) & i<n-p(n)-1 \\ \mathrm{Im} : H_i^{\phi}(X-X_o) \to H_i(X) & i=n-p(n)-1 \\ H_i(X) & i>n-p(n)-1 \end{cases}$$

Let q be the complementary perversity, so $q(n) = n-p(n)-2$. Similarly we have for the homology with compact supports

5.1.2

$$I_qH_i^c(X) = \begin{cases} H_i^c(X-X_o) & i<n-q(n)-1 \\ \mathrm{Im} : H_i^c(X-X_o) \to H_i^c(X) & i=n-q(n)-1 \\ H_i^c(X) & i>n-q(n)-1 \end{cases}$$

Note that for $i>1$, $H_i(X) \simeq H_i(X-X_o)$ (because $H_i(X_o) = 0$) and for $i\geq 1$, that $\mathrm{Im}(H_i^{\phi}(X-X_o) \to H_i(X)) = \mathrm{Im}(H_i^{\phi}(X-X_o) \to H_i(X-X_o))$ (resp. for $i>1$ $H_i^c(X) = H_i^{c'}(X-X_o)$ and for $i\geq 1$ $\mathrm{Im}(H_i^c(X-X_o) \to H_i^c(X)) = \mathrm{Im}(H_i^c(X-X_o) \to H_i^{c'}(X-X_o))$ where c' is the family of closed sets in $X-X_o$ which are relatively compact in X).

Alternatively, if we remove from X open conical neighbourhoods of each point of X_o, we get a manifold $\hat{X}$ with boundary $\partial\hat{X}$ the disjoint union of the links of the points of X_o. The inclusion of $\hat{X}$ in $X-X_o$ induces isomorphisms

$$H_i^c(\hat{X}) = H_i^c(X-X_o)$$
$$H_i(\hat{X}) = H_i^{\phi}(X-X_o)$$
$$H_i^c(\hat{X},\partial\hat{X}) = H_i^{c'}(X-X_o)$$

Hence $I_pH_i(X)$ can be expressed just in terms of the homology of the regular part $X-X_o$ with suitable families of supports. So if we denote by j the inclusion of $X-X_o$ in X, we can express 5.1.1 as

5.1.1'

$$H_i^{\phi}(X-X_o) \xrightarrow[\simeq]{j_*} I_pH_i(X) \qquad i<n-p(n)-1$$
$$H_i^{\phi}(X-X_o) \xrightarrow{j_*} I_pH_i(X) \xhookrightarrow{j^*} H_i(X-X_o) \qquad i=n-p(n)-1$$
$$I_pH_i(X) \xhookrightarrow[\simeq]{j^*} H_i(X-X_o) \qquad i>n-p(n)-1$$

and similarly 5.1.2 as

$$\begin{array}{lll}
H^c_i(X-X_o) \xrightarrow[\simeq]{j_*} I_qH^c_i(X) & & i<n-q(n)-1 \\
H^c_i(X-X_o) \xrightarrow{j_*} I_qH^c_i(X) & \xhookrightarrow{j^*} H^{c'}_i(X-X_o) & i=n-q(n)-1 \\
\phantom{H^c_i(X-X_o) \xrightarrow{j_*}} I_qH^c_i(X) & \xhookrightarrow[\simeq]{j^*} H^{c'}_i(X-X_o) & i>n-q(n)-1
\end{array}$$

All the properties of 4 can be checked directly using familiar properties of the homology of the regular part $X-X_o$. This exercise is very much recommended !

As an example, let us check Poincaré duality in the critical dimension $i=n-p(n)-1$ and $n-i=n-q(n)-1$, assuming X compact for simplicity. We assume that R is a field and that $X-X_o$ is R-oriented.

We have the diagram

$$\begin{array}{ccccccc}
H_i(X-X_o) & \times & H^c_{n-i}(X-X_o) & \xrightarrow{\cap} & H^c_o(X-X_o) & \searrow^{\varepsilon} & \\
\uparrow j^* & & \downarrow j_* & & & & \\
I_pH_i(X) & \times & I_qH_i(X) & \xrightarrow{\cap} & I_tH_o(X) & \xrightarrow{\varepsilon} & R \\
\uparrow j^* & & \downarrow j^* & & & \nearrow_{\varepsilon} & \\
H^c_i(X-X_o) & \times & H_{n-i}(X-X_o) & \xrightarrow{\cap} & H^c_o(X-X_o) & &
\end{array}$$

The middle line is a non degenerate pairing because the first and third lines define non degenerate pairings by Poincaré duality in $X-X_o$.

5.2 *Cone over a manifold.* Assume that L is a compact manifold of dimension n-1 and that X is the open cone $\overset{\circ}{c}L$ on L, namely the quotient of $L\times[0,\infty[$ by the equivalence relation which identifies $L\times\{0\}$ to a point (the vertex of the cone).

By 5.1.1 and 5.1.2, we have

$$I_pH_i(\overset{\circ}{c}L) = \begin{cases} 0 & i\leq n-p(n)-1 \\ H_{i-1}(L) & i>n-p(n)-1 \end{cases}$$

$$I_qH_i^c(\overset{\circ}{c}L) = \begin{cases} H_i(L) & i<n-q(n)-1 \\ 0 & i\geq n-q(n)-1 \end{cases}$$

5.3 *The Thom space of a vector bundle.* Let $\pi : E \longrightarrow V$ be an oriented real vector bundle of rank r over a compact manifold V of dimension m.

Let X be the Thom space of E : it is the one-point compactification $X = E \cup \{\infty\}$ of E. One can also describe it as follows.

Let us introduce a scalar product in the fibers of E and let B be the r-ball bundle over V formed by all vectors of E of norm ≤ 1. Its boundary S is a (r-1)-sphere bundle associated to E. Then $X = B \cup cS$ in the union of B with the cone over S. It is a pseudomanifold of dimension n = m+r with one singular point : the vertex $X_o = \{\infty\}$ of the cone.

Let $e \in H^r(V;\mathbb{Z})$ be the Euler class of E. We have the isomorphism :

$$H_i(E) \simeq H_{i-r}(V) \qquad \text{(Thom isomorphism)*}$$

$$H_i^c(E) \simeq H_i(V)$$

and the commutative diagram

$$\begin{array}{ccc} H_i^c(E) & \longrightarrow & H_i(E) \\ \wr\wr & & \wr\wr \\ H_i(V) & \xrightarrow{\cap e} & H_{i-r}(V) \end{array}$$

Hence by 5.1.2

5.3.1

$$I_pH_i(X) = \begin{cases} H_i(V) & i<n-p(n)-1 \\ \mathrm{Im} : H_i(V) \xrightarrow{\cap e} H_{i-r}(V) & i=n-p(n)-1 \\ H_{i-r}(V) & i>n-p(n)-1 \end{cases}$$

---

* If an element of $H_{i-r}(V)$ is represented by the geometric cycle ξ, then the corresponding element in $H_i(E)$ is represented by $\pi^{-1}(\xi)$.

since, for $i \geq 1$, $H_i(X) \cong H_i(X,\infty) = H_i(E) \cong H_{i-r}(V)$.

In particular, if $m=r$ and $p(n) = m-1$ (middle perversity), then

$$I_pH_i(X;\mathbb{Z}) = \begin{cases} H_i(V) & i<m \\ \tilde{e}\mathbb{Z} & i=m \\ H_{i-m}(V) & i>m \end{cases}$$

where $\tilde{e}$ denotes here the Euler number of the oriented bundle E, namely the evaluation of the Euler class of E over the fundamental class [V] of V (we have assumed V oriented and connected).

Using the diagram 5.1.3, we can check that the intersection pairing

$$I_pH_m(X) \times I_pH_m(X) \to \mathbb{Z}$$

maps (ea,eb) on eab. Hence one does not have Poincaré duality over $\mathbb{Z}$ (modulo torsion) if $|e| \neq 1$.

5.4 *The projective cone over a projective variety.* Let V be a smooth algebraic variety with $\dim_{\mathbb{C}} V = m-1$, embedded in $\mathbb{CP}^{N-1}$ considered as a hyperplane in $\mathbb{CP}^N$. Let X be the projective variety in $\mathbb{CP}^N$ union of the projective lines passing through a point $o \notin \mathbb{CP}^{N-1}$ and any point of V.

Topologically, X is the Thom space of the line bundle over V corresponding to an hyperplane section (equivalently the restriction to V of the normal bundle of $\mathbb{CP}^{N-1}$ in $\mathbb{CP}^N$).

For the middle perversity and rational coefficients, we have by 5.3.1

$$IH_i(X) = \begin{cases} H_i(V) & i \leq m \\ H_{i-2}(V) & i>m \end{cases}$$

because in dimension m, the cap product with the Euler class is the map $H_m(V) \to H_{m-2}(V)$ given by intersection with a hyperplane section H which is an isomorphism by the Hard Lefschetz theorem for V (cf. Griffiths-Harris, [5]).

For $i>m$, the isomorphism $IH_i(X) \longrightarrow H_{i-2}(V)$ is induced by intersection with the hyperplane CP^{N-1}.

In that particular case, it is easy to check the *hard Lefschetz theorem for X*, namely that the map

$$\cap H^k : IH_{m+k}(X) \overset{\simeq}{\rightarrow} IH_{m-k}(X)$$

obtained by iterating k times the map H induced by intersection with a hyperplane is an isomorphism (it is well defined because a generic hyperplane section is of perversity $\bar{0}$), the coefficients being the rational numbers.

This follows from the commutativity of the diagram

$$\begin{array}{ccc} H_{m+k-2}(V) & \xrightarrow{\cap H^{k-1}} & H_{m-k}(V) \\ \uparrow \wr & & \downarrow \wr \\ IH_{m+k}(X) & \xrightarrow{\cap H^{k}} & IH_{m-k}(X) \end{array}$$

and the fact that the top horizontal map is an isomorphism (by the hard Lefschetz theorem for the smooth algebraic variety V).

Note also that the link of the singular point of X is the circle bundle S associated to the line bundle η. So according to 5.2, if U is an open conical neighbourhood of the singular point of X, we have

$$IH_i(U) = \begin{cases} 0 & i\leq m \\ H_{i-1}(X) & i>m \end{cases}$$

The Thom-Gysin exact sequence of the circle bundle $S \rightarrow V$ gives for $i>m+1$ the short exact sequence :

$$0 \rightarrow H_i(V) \xrightarrow{\cap H} H_{i-2}(V) \rightarrow H_{i-1}(S) \rightarrow 0$$

because for $i>m-1$ the map $\cap H$ is injective (again by the Lefschetz theorem). A complement of the image of $\cap H$ is the subspace P_{i-2} of $H_{i-2}(V)$ made up of the Poincaré dual of the primitive elements. Recall that the space P^k of primitive elements of degree k, where $k \le \dim_{\mathbb{C}} V = m-1$, is the subspace of $H^k(V;\mathbb{Q})$ formed by the elements α such that $L^{n-k}\alpha = 0$, where L is the map induced by cup product with the class dual to a hyperplane section. By the theorem of Lefschetz,

$$H^r(V,\mathbb{Q}) = \oplus L^s P^{r-2s}$$

hence

$$\oplus H^r(V,\mathbb{Q}) = \oplus P^k \oplus \mathrm{ImL}.$$

Hence we get

$$IH_i(U) = \begin{cases} 0 & i \le m \\ P_{i-2} & i > m \end{cases}$$

REFERENCES

[1] A.Borel and J.C.Moore : "Homology theory for locally compact spaces". Michigan Math. J., 7 (1960), 137-159.

[2] J. Dixmier : "Homologie et cohomologie singulières". Séminaire Henri Cartan, le année, 1948/1949, "Topologie algébrique", exposé No 5, 2e éd., Paris, Ecole normale supérieure, 1955.

[3] M. Goresky and R. MacPherson : "Intersection homology theory". Topology, 19 (1980), 135-162.

[4] M. Goresky and R. MacPherson : "Intersection homology II". Invent. Math., 72 (1983), 77-129.

[5] P. Griffiths and J. Harris : "Principles of algebraic geometry", New York, J.Wiley, 1978.

[6] J.F.P. Hudson : "Piecewise linear topology", New York, W.A. Benjamin, 1969, Mathematics lecture note series.

[7] S. Lojasiewicz : "Triangulation of semi-analytic sets". Ann. Scuola Norm. Sup. Pisa (3), 18 (1964), 449-474.

[8] C. MacCrory : "Stratified general position", in "Algebraic and geometric topology", (Proc., Symp., Santa Barbara, in honor of Raymond L. Wilder, 1977), ed. by K.C. Millett, Berlin, Springer, 1978, Lecture notes in mathematics, No 664, 142-146.

[9] J. Mather : "Stratifications and mappings", in "Dynamical systems", (Proc.,Symp., Univ., Bahia, Salvador, 1971), ed. by M.M. Pcixoto, New York, Academic press, 1973, 195-232.

[10] R. Thom : "Espaces fibrés en sphères et carrés de Steenrod". Ann. Sci. Ecole Norm. Sup., Paris, (3), 69 (1952), 109-182.

Intersection Cohomology
Progress in Mathematics, Vol. 50
© 1984 Birkhäuser Boston, Inc.

II. FROM PL TO SHEAF THEORY

(Rock to Bach)

Notes by Nathan Habegger

§ 0 INTRODUCTION

We modern mathematicians often tend to hide the primitive beginnings of our knowledge, much to the detriment of beginning students. Of course, we are each entitled to have our own preference in music. But let's not forget that to the original geometers, homology was not a derived functor on the category of sheaves, nor a functor from spaces to groups satisfying axioms 1-7. Homology was space itself, space with multiplicities.

It seems fitting that intersection homology was born out of the study of the geometry of cycles themselves. (M. Goresky and R. MacPherson attribute their motivation to questions posed by D. Sullivan). However there was perhaps a danger of the subject remaining little more than a curiosity, in spite of the fact that these invariants are among the few known which are not homotopy theoretic in nature.

Enter in sheaf theory, whose theoretical power lies in the passage from the local to the global. With the introduction of Deligne's sheaf and the subsequent axiomatic development, intersection homology had matured.

The intent of these notes is to take the reader through this evolutionary process, beginning with the pl chain theoretic definitions (see I) and ending with the sheaf theoretic definitions. Admittedly, nothing substancial is contained herein, given that the sheaf theoretic development may be done independently of the pl theory. On the other hand, the pl theory is our intuitive guide, and it is consoling to know that, yes, the two theories are the same.

The material for these notes was taken from § 2 and § 3 of Intersection Homology II. I have also profited from an introductory lecture by Borel and from discussions with Haefliger. Thanks also to L. Kaup who took the time to make helpful suggestions for the improvement of an earlier version of these notes.

§ 1 THE CALCULUS OF CHAINS

We will assume that the reader is familiar with the language of sheaf theory and with pl theory (as outlined in I). All spaces will be pl and paracompact. Homology will be taken to be Borel-Moore homology with coefficients in a *fixed* ring R with unit (or, what amounts to the same, singular homology, using locally finite infinite chains).

The local homology sheaves, $\mathcal{H}_i$, are generated by the presheaves $U \mapsto H_i(U)$. However, if X is of dimension n, the presheaf $U \mapsto H_n(U)$ is actually a sheaf. In particular, elements of $H_n(X)$ are just global sections of the sheaf $\mathcal{H}_n$, i.e. $H_n(X) = \mathcal{H}_n(X)$. They have support (a closed pl subspace) and local values (compatible with the topology of $\mathcal{H}_n$). Thus a homology class is "space" with "multiplicities". Notice that on the regular part (e.g. on interiors of n-dimensional simplices) the sheaf $\mathcal{H}_n$ is locally isomorphic to the constant sheaf R.

If A and B are closed pl subspaces of X of dimension i and i-1 respectively, then chains in $C_i(X)$ which satisfy $|\xi| \subset A$ and $|\partial\xi| \subset B$, correspond bijectively to elements $[\xi] \in H_i(A,B)$: the chain ξ is a cycle in the quotient complex $\frac{C.(A)}{C.(B)}$ whose homology is $H.(A,B)$.

Thus in order to prescribe chains, we need only describe sets and homology classes. Moreover, the isomorphism $H_i(A,B) = H_i(A\backslash B)$ shows that the classes need only be described *locally*, that is as sections of the homology sheaf $\mathcal{H}_i$ of $A\backslash B$.

For example, the chain ξ is completely described by $|\xi|$, its support, $|\partial\xi|$ and $[\xi] \in H_i(|\xi|,|\partial\xi|)$. Moreover $[\xi]$ is just a continuous family $[\xi]_x \in (\mathcal{H}_i)_x$ of non-zero local homology classes on

$|\xi|\setminus|\partial\xi|$. Such a description of a chain will be referred to as *the local description.*

Products of chains $\xi\times\eta$, cones on compact chains $c\eta$, direct image under proper pl maps $f_*\eta$, may all be defined by describing appropriate sets and homology classes. For example, the chain $c\eta$ is defined by the class $\partial_*^{-1}[\eta]$ where $\partial_* : H_{i+1}(c|\eta|, |\eta| \cup c|\partial\eta|) \tilde{\to} H_i(|\eta|, |\partial\eta|)$. (So $\partial c\eta = \eta - c\partial\eta$).

One may describe chains using oriented simplices and multiplicities : If $\xi \in C_i(X)$ and $|\xi|$ is a subcomplex of some triangulation T of X, then $\xi = r_j\Delta_j$ where Δ_j are the oriented i-simplices and $r_j \in R$. If F is a closed pl subset of X which is also a subcomplex of T, then the restriction of ξ to F, denoted by $\xi \cap F$, is given by $\sum_{\Delta_j \subset F} r_j\Delta_j$. Note that $\xi \cap F$ is in $C_i(F)$ and also in $C_i(X)$, since $C_i(F) \subset C_i(X)$. Note also that if $|\xi| \cap F$ has dimension less than i, then $\xi \cap F = 0$, the empty chain. In general $\xi \cap (F_1 \cup F_2) = \xi \cap F_1 + \xi \cap F_2 - \xi \cap (F_1 \cap F_2)$. Note that $\partial\xi \cap F \neq \partial(\xi \cap F)$, in general.

The restriction of a chain ξ to an open set U, $\xi \cap U \in C_i(U)$, is perhaps best described as that chain whose local description is the same (in U) as ξ. It follows that $|\xi \cap U| = |\xi| \cap U$ and $\partial(\xi \cap U) = \partial\xi \cap U$. Alternatively, one may give a description of $\xi \cap U$ using simplices (cf. I § 2).

§ 2 INTERSECTION HOMOLOGY OF A PRODUCT WITH EUCLIDEAN SPACE

Let X^n be a (paracompact) stratified pl pseudomanifold. Let $\mathbb{R}\times X$ denote the pseudomanifold with stratification $(\mathbb{R}\times X)_{i+1} = \mathbb{R}\times X_i$. The correspondance $\xi \to \mathbb{R}\times\xi$ gives a map of complexes $IC_\cdot(X) \xrightarrow{S} IC_{\cdot+1}(\mathbb{R}\times X)$ called suspension.

PROPOSITION 2.1 : *The suspension map induces an isomorphism on homology*

$$IH_i(X) \tilde{\to} IH_{i+1}(\mathbb{R}\times X).$$

In order to prove this proposition, we need the following

LEMMA 2.2 : *Let* $\xi \in IC_i(\mathbb{R}\times X)$ *be a cycle with support in* $\mathbb{R}_+\times X$. *Then* $\xi = \partial\eta$ *for a chain* $\eta \in IC_{i+1}(\mathbb{R}\times X)$ *supported in* $\mathbb{R}_+\times X$.

Proof : Let $p_1 : \mathbb{R}\times X \to \mathbb{R}$, $p_2 : \mathbb{R}\times X \to X$ be the projections. Define a map

$$f : \mathbb{R}_+\times|\xi| \to \mathbb{R}\times X \quad \text{by}$$
$$(t,x) \to (t+p_1(x),\ p_2(x)).$$

The chain $f_*(\mathbb{R}_+\times\xi)$ has boundary ξ and is in $IC_{i+1}(\mathbb{R}\times X)$, since its support, contained in $f(\mathbb{R}_+\times|\xi|)$, satisfies the intersection conditions. (See diagram).

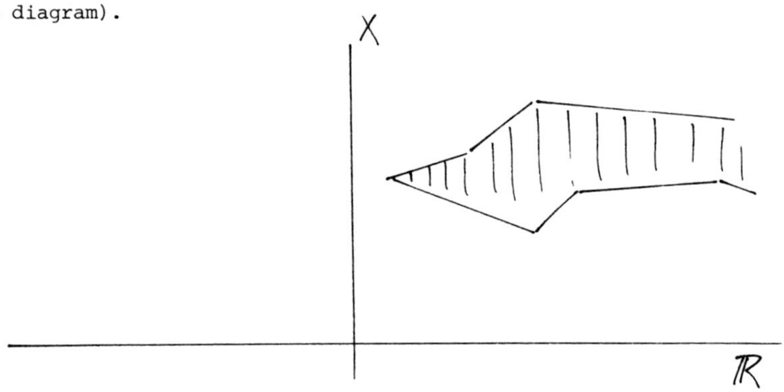

2.3 A homology of a cycle to infinity. (cf.2.2)

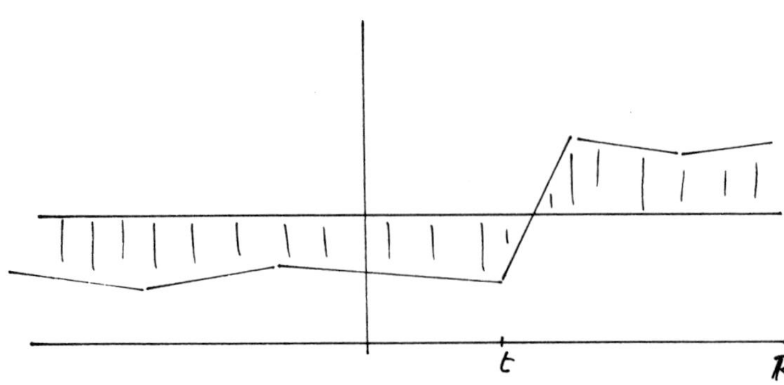

2.4 A homology to a suspended cycle. (cf.2.1)

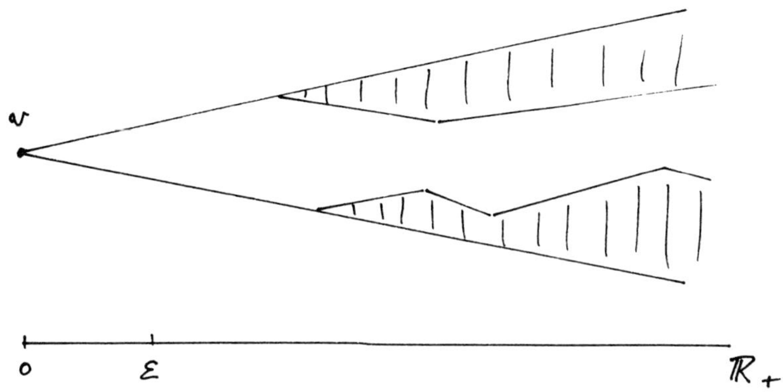

2.5 A homology to a conical cycle. (cf.3.1)

Proof of 2.1 : Since S is injective, it suffices to show that the quotient complex is acyclic, i.e. we must show that if $\xi \in IC_i(\mathbb{R}\times X)$ has $\partial\xi = \mathbb{R}\times\eta, \eta \in IC_{i-2}(X)$, then $\xi = \partial\mu + \mathbb{R}\times\gamma$ where $\mu \in IC_{i+1}(\mathbb{R}\times X)$ and $\gamma \in IC_{i-1}(X)$.

Briefly, the idea is as follows : Cut the chain ξ transversally, to obtain γ , and use the lemma to find a homology of ξ to $\mathbb{R}\times\gamma$. (See diagram). In detail :

Let $p_1 : \mathbb{R}\times X \to \mathbb{R}$ be the projection. Let T be a triangulation of $\mathbb{R}\times X$ for which $|\xi|$, $(\mathbb{R}\times X)_j$ are subcomplexes.

If X is connected (otherwise we argue componentwise), then T has only countably many vertices, and hence we may find $t \in \mathbb{R}$ such that $p_1^{-1}(\{t\})$ contains no vertex. So $|\xi|$ intersects $\{t\}\times X$ transversely.

Thus $\xi = \xi_+ + \xi_-$ where $\xi_+ = \xi \cap [t,\infty]\times X$ and $\xi_- = \xi \cap (-\infty,t]\times X$. (N.B. $\xi \cap \{t\}\times X = 0$, for dimensional reasons).

Then $\partial\xi_+ = [t,\infty)\times\eta + \{t\}\times\gamma$ where $\gamma \in IC_{i-1}(X)$ and $\partial\gamma = \eta$. It follows that $\xi_+ + [t,\infty)\times\gamma$ is a cycle in $IC_i(\mathbb{R}\times X)$ with support in $[t,\infty)\times X$, so by the lemma, it bounds. Similarly $\xi_- + (-\infty,t]\times\gamma$ bounds, so $\xi + \mathbb{R}\times\gamma =$
$= (\xi_+ + [t,\infty)\times\gamma) + (\xi_- + (-\infty,t]\times\gamma)$ also bounds.

§ 3 INTERSECTION HOMOLOGY OF A CONE

Let L^{k-1} be a compact stratified pl pseudomanifold. Let $\overset{\circ}{c}L$ denote the (open) cone with vertex v and stratification

$$(\overset{\circ}{c}L)_i = \begin{cases} \overset{\circ}{c}L_{i-1} & i>0 \\ \{v\} & i=0 \end{cases}$$

Problem : If $\xi \in IC_{i-1}(L)$, when is $\overset{\circ}{c}\xi \in IC_i(\overset{\circ}{c}L)$.

Answer : For $i>k-p(k)$, any ξ
$i=k-p(k)$, only cycles
$i<k-p(k)$, no ξ.

Proof : The conditions on intersection with the strata other than the vertex v are automatically satisfied. Checking the conditions on intersection with the vertex v yields the restrictions above.

Q.E.D.

Notation : For a chain complex $C_{\cdot}$, $\tau_{\cdot\geq r}C_{\cdot}$ denotes the "truncated complex" defined by

$$(\tau_{\cdot\geq r}C_{\cdot})_i = \begin{cases} C_i & i>r \\ \ker C_r \xrightarrow{\partial} C_{r-1} & i=r \\ 0 & i<r. \end{cases}$$

The answer above implies that we have a map of complexes (called coning)

$$\tau_{\cdot\geq k-p(k)} IC_{\cdot-1}(L) \xrightarrow{\overset{\circ}{c}} IC_{\cdot}(\overset{\circ}{c}L)$$

given by

$$\xi \to \overset{\circ}{c}\xi.$$

PROPOSITION 3.1 : *Coning induces an isomorphism on homology, i.e.*

$$IH_i(\overset{\circ}{c}L) = \begin{cases} IH_{i-1}(L) & i \geq k-p(k) \\ 0 & i < k-p(k) \end{cases}$$

Proof of 3.1 : The map $\overset{\circ}{c}$ is injective. So it suffices to show that if $\xi \in IC_i(\overset{\circ}{c}L)$ and $\partial\xi = \overset{\circ}{c}\eta$, $\eta \in IC_{i-2}(L)$, then $\xi = \partial\mu + \overset{\circ}{c}\gamma$, $\mu \in IC_{i+1}(\overset{\circ}{c}L)$, $\gamma \in IC_{i-1}(L)$.

Let $\pi : \overset{\circ}{c}L \to \mathbb{R}_+$ be the projection and let $N_\varepsilon = \pi^{-1}[0,\varepsilon]$ be the closed ε-neighborhood of the cone vertex v. Choose ε sufficiety small so that $N_\varepsilon \cap |\xi|$ contains no vertex other than v of some triangulation of $|\xi|$. Then $\xi \cap N_\varepsilon$ is conical, i.e. $\xi \cap N_\varepsilon = \overset{\circ}{c}\gamma \cap N_\varepsilon$.

It follows that $\gamma \in IC_{i-1}(L)$ and that $\partial\gamma = -\eta$. So $\xi - \overset{\circ}{c}\gamma$ is a cycle with support in $\pi^{-1}[\varepsilon,\infty) \overset{PL}{\cong} \mathbb{R}_+ \times X$. By lemma 2.2, $\xi - \overset{\circ}{c}\gamma$ is a boundary. (See diagram). Q.E.D.

If we remove the cone vertex v from $\overset{\circ}{c}L$, we have an isomorphism $\overset{\circ}{c}L\backslash\{v\} \overset{PL}{\cong} \mathbb{R}\times L$, however the obvious map $\overset{\circ}{c}L\backslash\{v\} \to L$ is not pl (the standard mistake). So the following proposition is (technically) not a consequence of 2.1,

PROPOSITION 3.2 *The map* $IC_{\cdot-1}(L) \xrightarrow{\overset{\circ}{c}\backslash v} IC_\cdot(\overset{\circ}{c}L\backslash\{v\})$ *given by* $\xi \to \overset{\circ}{c}\xi\backslash v = \overset{\circ}{c}\xi \cap (\overset{\circ}{c}L\backslash\{v\})$ *induces an isomorphism on homology.*

Proof : The proof is similar to that of 2.1 using the projection $\overset{\circ}{c}L \to \mathbb{R}_+$ (which is pl) in place of the projection $\mathbb{R}\times X \to \mathbb{R}$.

§ 4 THE LOCAL INTERSECTION HOMOLOGY GROUPS

PROPOSITION 4.1 *Let* L^{k-1} *be a compact stratified pseudomanifold. There is a commutative diagram*

$$\begin{array}{ccc} \tau_{\cdot \geq n-p(k)} IC_{\cdot-(n-k+1)}(L) & \hookrightarrow & IC_{\cdot-(n-k+1)}(L) \\ \downarrow & & \downarrow \\ IC_\cdot(\mathbb{R}^{n-k} \times \overset{\circ}{c}L) & \xrightarrow{res} & IC_\cdot(\mathbb{R}^{n-k} \times (\overset{\circ}{c}L\backslash\{v\})) \end{array}$$

with vertical arrows inducing isomorphisms on homology.

In particular, $IH_i(\mathbb{R}^{n-k}\,\overset{\circ}{c}L) = \begin{cases} IH_{i-(n-k+1)}(L) & i \geq n-p(k) \\ 0 & i < n-p(k) \end{cases}$

Proof : The left vertical arrow is the map $\xi \to \mathbb{R}^{n-k}\times\overset{\circ}{c}\xi$ which is the composite $\underbrace{S \circ S \circ S}_{n-k} \circ \overset{\circ}{c}$ so it induces an isomorphism by 2.1 and 3.1.

The right vertical arrow is the map $\xi \to \mathbb{R}^{n-k}\times\overset{\circ}{c}\xi\backslash v$ which is the composite $\underbrace{S \circ S \qquad S}_{n-k} \circ \overset{\circ}{c}\backslash v$ and so induces an isomorphism by 2.1 and 3.2.

Commutativity holds, since suspension commutes with restriction. Q.E.D.

Restriction homomorphisms are functorial for inclusions of open sets (use the local description of chains). So the correspondance $U \to IC_i(U)$ defines a presheaf. The axioms for a sheaf are satisfied : If $\xi_k \in IC_i(U_k)$ are a compatible family of chains, their union $\cup\xi_k \in IC_i(\cup U_k)$ is a uniquely defined chain (local description again).

This sheaf will be denoted by $\mathcal{IC}_i$. When we wish to consider positively graded complexes of sheaves with differential of degree +1, we will use *codimension* as variable and use the notation $\mathcal{IC}_{n-\cdot}$ (where $\cdot$ varies from 0 to n).

For a complex $F^{\cdot}$ of sheaves $\mathcal{H}^j(F^{\cdot})$ denotes the jth derived sheaf. $\mathcal{H}^j(F^{\cdot})_x$ is the local homology at x.

PROPOSITION 4.2 : *Let* $i : \mathbb{R}^{n-k}\times(\overset{\circ}{c}L\backslash\{v\}) \to \mathbb{R}^{n-k}\times\overset{\circ}{c}L$ *denote the inclusion. Let* $x \in \mathbb{R}^{n-k}\times\{v\}$.

Then the maps

$$\mathcal{IC}_{\cdot}(\mathbb{R}^{n-k}\times\overset{\circ}{c}L) \to (\mathcal{IC}_{\cdot})_x$$
$$\mathcal{IC}_{\cdot}(\mathbb{R}^{n-k}\times(\overset{\circ}{c}L\backslash\{v\})) \to (i_*\mathcal{IC}_{\cdot})_x$$

induce isomorphisms on homology.

COROLLARY 4.3 : *For* $x \in \mathbb{R}^{n-k}\times\{v\}$, *i as above, then*

$$\mathcal{H}^j(\mathcal{IC}_{n-\cdot})_x = \begin{cases} \mathcal{H}^j(i_*\mathcal{IC}_{n-\cdot})_x & \text{if } j \leq p(k) \\ 0 & \text{if } j > p(k) \end{cases}$$

The sheaves $\mathcal{H}^j(IC_{n-\cdot})$, $\mathcal{H}^j(i_*IC_{n-\cdot})$ *are constant on* $\mathbb{R}^{n-k}\times\{v\}$.

Proof of 4.2 : For $x\in\mathbb{R}^{n-k}\times\{v\}$, there are arbitrarily small neighborhoods V_ε of x and pl isomorphisms $i_\varepsilon : \mathbb{R}^{n-k}\times \overset{\circ}{c}L \xrightarrow[\simeq]{PL} V_\varepsilon$ preserving product and cone structure (and hence stratified structure). For these neighborhoods the following diagram commutes

$$\begin{array}{ccc} \tau_{\cdot\geq n-p(k)}IC_{\cdot-(n-k+1)}(L) & \longrightarrow & IC_\cdot(\mathbb{R}^{n-k}\times\overset{\circ}{c}L) \\ \downarrow & & \downarrow {\scriptstyle i_{\varepsilon_*}} \\ IC_\cdot(\mathbb{R}^{n-k}\times\overset{\circ}{c}L) & \xrightarrow{\text{res}} & IC_\cdot(V_\varepsilon) \end{array}$$

It follows that the restriction map induces an isomorphism on homology. Passing to the limit over such neighborhoods yields the first assertion. The second is proved similarly.

Proof of 4.3 Combining 4.1 and 4.2 we obtain a commutative diagram

$$\begin{array}{ccc} \tau_{\cdot\leq p(k)}IC_{k-1-\cdot}(L) & \longrightarrow & IC_{k-1-\cdot}(L) \\ \downarrow & & \downarrow \\ (IC_{n-\cdot})_x & \longrightarrow & (i_*IC_{n-\cdot})_x \end{array}$$

with vertical arrows inducing isomorphisms on homology. The first assertion follows.

To see that the sheaves are constant, we make use of 4.2 and the following :

Let F be a presheaf and $\tilde{F}$ the sheaf it generates. The map $X\times F(X)\to\tilde{F}$, which takes (x,s) to the germ of s at X, is a local homeomorphism, where F(X) has the discrete topology. Let S be a subset of X for which $F(X)\to\tilde{F}_x$ is a bijection, $x\in S$. Then $S\times F(X)\to\tilde{F}|_S$ is an isomorphism of sheaves. It follows that if $\mathcal{F}^\cdot$ is a complex of sheaves such that $\mathcal{F}^\cdot(X)\to\mathcal{F}^\cdot_x$ induces isomorphisms on homology, $x\in S$, then the map $S\times H^j(\mathcal{F}^\cdot(X))\to\mathcal{H}^j(\mathcal{F}^\cdot)|_S$ is an isomorphism of sheaves.

§ 5. THE SHEAVES $\mathcal{IC}_{\cdot}$ ARE SOFT

In the following, we will restrict attention to *positively graded* complexes of sheaves (i.e. $F^i = 0$, $i<0$). A map $F^{\cdot} \to G^{\cdot}$ is a *quasi isomorphism* (or *resolution)* if it induces an isomorphism $H^i(F) \overset{\simeq}{\to} H^i(G^{\cdot})$ of derived sheaves for all i.

The *hypercohomology* groups are defined by $\mathbb{H}^i(X;F^{\cdot}) \equiv H^i(I^{\cdot}(X))$, where $F^{\cdot} \to I^{\cdot}$ is any injective resolution. From the spectral sequence $H^p(X;H^q(F^{\cdot})) \Rightarrow \mathbb{H}^{p+q}(X;F^{\cdot})$ [Godement, page 178] we see that a quasi isomorphism induces an isomorphism on hypercohomology.

From the spectral sequence

$$H^p(H^q(X;F^{\cdot})) \Rightarrow \mathbb{H}^{p+q}(X;F^{\cdot})$$

we see that if each F^i is *cohomologically trivial* (i.e. $H^q(X;F^i) = 0$, $q>0$), then $\mathbb{H}^i(X;F^{\cdot}) = H^i(F^{\cdot}(X))$. Examples of cohomologically trivial sheaves include injective sheaves, flabby sheaves and (for paracompact spaces) fine sheaves and soft sheaves.

Using the above and the fact that the direct image of an injective sheaf is injective, it is an exercise to show the following :

Let $f : X \to Y$. If F is cohomologically trivial on *all open* subsets of X, then f_*F is cohomologically trivial on all open sets of Y. If $F^{\cdot} \to G^{\cdot}$ is a quasi isomorphism, then $f_*F^{\cdot} \to f_*G^{\cdot}$ is a quasi isomorphism provided that F^i, G^i are cohomologically trivial on all open sets of X. In particular, if $Rf_*F^{\cdot}$ *denotes* $f_*I^{\cdot}$, where $F^{\cdot} \to I^{\cdot}$ is an injective resolution, then $f_*F^{\cdot} \to Rf_*F^{\cdot}$ is a quasi isomorphism, provided each F^i is cohomologically trivial on all open sets of X.

The above serves as motivation for the following proposition and as proof of its corollaries :

PROPOSITION 5.1 : *The sheaves* $\mathcal{IC}_i$ *are soft.*

COROLLARY 5.2 : $\mathbb{H}^i(X;\mathcal{IC}_{n-\cdot}) = IH_{n-i}(X)$.

COROLLARY 5.3 : *Let* $i_k : U_k \to U_{k+1}$ *denote the inclusion. Let* $U_j = X-X_{n-j}$. *Then* $i_{k*}\mathcal{IC}_{n-\cdot}|_{U_k} \to Ri_{k*}\mathcal{IC}_{n-\cdot}|_{U_k}$ *is a quasi isomorphism.*

[Note : In their paper, Goresky and MacPherson claim the intersection sheaves are fine (which implies soft). The ordinary sheaves C_i are fine, since they are modules over the sheaf of constructible functions (i.e. functions constant on interiors of simplices), which is obviously fine. However, the sheaves IC_i are *not* modules over the sheaf of constructible functions : an intersection chain cannot be broken into pieces arbitrarily ! It may be that the sheaves are fine, but the point is rather moot, since 5.1 and its corollaries are sufficient for our purposes.]

Let T be a triangulation of X and T' the first barycentric subdivision. If v is a vertex of T, we denote by N'(v) the (closed) neighborhood of v, union of all closed simplices of T' containing v as a vertex.

Remark : If $\xi \in IC_i(X)$ and T is any triangulation of X for which $|\xi|$, X_j are subcomplexes, then $\xi \cap N'(v)$ is in $IC_i(X)$.

Proof of 5.1 : We must show : If F is a closed set of X, then $IC_i(X) \to IC_i(F)$ is surjective. Let $s \in IC_i(F)$ be given. s is represented by $\xi \in IC_i(U)$ for some open neighborhood U of F [Godement, page 150] Let $N'(F) = \cup N'(v)$ where the union is taken over all vertices (of a triangulation T *of U*) belonging to a (closed) simplex intersecting F.

By taking the triangulation T to be sufficiently fine, we may assume N'(F) is a *closed subset of X*. Assume $|\xi|$ and $U \cap X_j$ are subcomplexes of T. Then, by the remark, $\xi \cap N'(F)$ $(= \Sigma(\xi \cap N'(v)))$ is in $IC_i(X)$ (its support is closed in X) and restricts to s.

§ 6 AXIOMS FOR THE INTERSECTION HOMOLOGY SHEAVES

Let X be a stratified pl pseudomanifold. Put $U_k = X \setminus X_{n-k}$ and let $i_k : U_k \to U_{k+1}$ denote the inclusion. $S_{n-k} = X_{n-k} \setminus X_{n-k-1}$ denotes the n-k dimensional stratum of X. U_2 is an open dense subset of X and is a manifold.

The following theorem is the starting point of the sheaf theoretic axiomatic development of intersection homology (cf. V, §2,3).

THEOREM 6.1 : *The complex* $IC_{n-\cdot}$ *of sheaves satisfies the following :*

1) $IC_{n-\cdot}$ *is a bounded complex, zero in negative degrees.* $IC_{n-\cdot}|_{U_2}$ *is quasi isomorphic to the orientation sheaf of* U_2.

2) $\mathcal{H}^j(IC_{n-\cdot})_x = 0$ *for* $j>p(k)$, $x\in S_{n-k}$.

3) *The attaching map* $IC_{n-\cdot}|_{U_{k+1}} \to Ri_{k*}IC_{n-\cdot}|_{U_k}$ *induces an isomorphism*

$$\mathcal{H}^j(IC_{n-\cdot})_x \to \mathcal{H}^j(Ri_{k*}IC_{n-\cdot})_x \text{ for } x\in S_{n-k},\ j\leq p(k).$$

Proof : The first assertion in 1) is obvious. For $x\in U_2$, there are no intersection conditions near x, so

$IC_{n-\cdot}|_{U_2} = C_{n-\cdot}|_{U_2}$ and hence $\mathcal{H}^i(IC_{n-\cdot})_x = \begin{cases} 0 & i>0 \\ R & i=0 \end{cases}$, the identification $\mathcal{H}^0(IC_{n-\cdot})_x = R$ depending on the local orientation.
So $\mathcal{H}^0(IC_{n-\cdot})|_{U_2}$ is the orientation sheaf of U_2 and $\mathcal{H}^0(IC_{n-\cdot}|_{U_2} \to IC_{n-\cdot}|_{U_2})$ is a quasi isomorphism.

Since the statements in 2) and 3) are local, and since x has a neighborhood parametrized by $\mathbb{R}^{n-k}\times \overset{\circ}{c}L$, Corollary 4.3 applies. (We have used 5.3 to replace $Ri_{k*}IC_{n-\cdot}$ by $i_{k*}IC_{n-\cdot}$).

Intersection Cohomology
Progress in Mathematics, Vol. 50
© 1984 Birkhäuser Boston, Inc.

III. A SAMPLE COMPUTATION OF INTERSECTION HOMOLOGY

by M. Goresky and R. MacPherson

In this example we compute the intersection homology of the Cartesian product of S^1 with the suspension of the 3-torus. We use the notation $X = (\Sigma T^3) \times S^1$. By choosing a basepoint $\{p\}$ in S^1 we can identify the following cycles in T^3 :

$$T_a^1 = S^1 \times \{p\} \times \{p\};\ T_b^1 = \{p\} \times S^1 \times \{p\};\ T_c^1 = \{p\} \times \{p\} \times S^1$$

$$T_a^2 = \{p\} \times S^1 \times S^1 \ ;\ T_b^2 = S^1 \times \{p\} \times S^1 \ ;\ T_c^2 = S^1 \times S^1 \times \{p\} \ .$$

The space X is stratified in the obvious way with one stratum of dimension 5 and two singular strata of dimension 1 . Thus X has even codimension singularities and it is normal since the link of the singular stratum is T^3 which is connected. The intersection homology groups with perversity $\bar{p}$ are determined by the single number $p = \bar{p}(4)$ which can be 0, 1 or 2 . Thus, a chain ξ is allowable in $IC_i^p(X)$ if it intersects the singular set in dimension $\leq i - 4 + p$ and its boundary intersects the singular set in dimension $\leq i - 5 + p$.

In the following table we give a list of allowable cycles which generate the intersection homology groups of given dimension and perversity.

For example, the 3-cycle $(\Sigma T_a^1) \times S^1$ is allowable in IC_3^2 but is not allowable in IC_3^0 or IC_3^1 since it intersects the singular set in dimension 1 . Similarly the 3-cycle $T_a^2 \times S^1$ generates a nonzero class in $IH_3^0(X)$ but it is 0 in $IH_3^1(X)$ since it is the boundary of $\text{cone}(T_a^2) \times S^1$ which is an allowable chain in IC_4^1 .

	p = 0	p = 1	p = 2
IH_5	$\Sigma T^3 \times S^1$	$\Sigma T^3 \times S^1$	$\Sigma T^3 \times S^1$
IH_4	ΣT^3	ΣT^3 $\Sigma T_a^2 \times S^1$ $\Sigma T_b^2 \times S^1$ $\Sigma T_c^2 \times S^1$	ΣT^3 $\Sigma T_a^2 \times S^1$ $\Sigma T_b^2 \times S^1$ $\Sigma T_c^2 \times S^1$
IH_3	$T_a^2 \times S^1$ $T_b^2 \times S^1$ $T_c^2 \times S^1$	ΣT_a^2 ΣT_b^2 ΣT_c^2	ΣT_a^2 $\Sigma T_a^1 \times S^1$ ΣT_b^2 $\Sigma T_b^1 \times S^1$ ΣT_c^2 $\Sigma T_c^1 \times S^1$
IH_2	T_a^2 $T_a^1 \times S^1$ T_b^2 $T_b^1 \times S^1$ T_c^2 $T_c^1 \times S^1$	$T_a^1 \times S^1$ $T_b^1 \times S^1$ $T_c^1 \times S^1$	ΣT_a^1 ΣT_b^1 ΣT_c^1
IH_1	T_a^1 T_b^1 $pt \times S^1$ T_c^1	T_a^1 T_b^1 $pt \times S^1$ T_c^1	$pt \times S^1$
IH_0	pt	pt	pt

By reading across this table we can recover the Poincaré duality map $H^{5-i}(X) = IH_i^0(X) \to H_i(X)$ which is an isomorphism for $i = 0$ and 5, an injection for $i = 4$, a surjection for $i = 1$, and is zero for $i = 2$ and 3 . We also observe that the cohomology Betti numbers (1,4,6,3,1,1) are not palindromic whereas the intersection homology Betti numbers ($\bar{p} = 1$) are palindromic : (1,4,3,3,4,1). In fact the nongenerate pairing $IH_i^1 \times IH_{5-i}^1 \to Z$ is geometric. For example, the cycle $\Sigma T_a^2 \times S^1$ in IH_4^1 is dual to the cycle T_a^1 in IH_1^1 , and intersects it transversally at a single point. The paring between IH_i^0 and IH_{5-i}^2 is similar.

We verify the impossibility of constructing a product in homology because the cycles $\Sigma T_a^2 \times S^1$ and ΣT_a^1 are transverse, but they intersect in $(\Sigma pt) \times pt$ which is not a cycle. The cycles $\Sigma T_a^2 \times S^1$ and ΣT_b^2 do not come from cohomology but they are intersectable since they live in IH^1. Their product is ΣT_c^1 .

This local intersection homology at a point x in the singularity set is easily calculated since a cycle represents 0 in $IH_i^p(X,X-x)$ unless it contains a whole neighborhood of x in the singular stratum. This gives the following table from which we can verify the "support condition": the local homology sheaf associated to IH_i vanishes for all $i < 5 - p$.

	p = 0	p = 1	p = 2
$IH_5(X,X-x)$	Z	Z	Z
$IH_4(X,X-x)$	0	Z ⊕ Z ⊕ Z	Z ⊕ Z ⊕ Z
$IH_3(X,X-x)$	0	0	Z ⊕ Z ⊕ Z
$IH_2(X,X-x)$	0	0	0

Exercise: Compute a similar table for the local compact support homology at a point in the singular set.

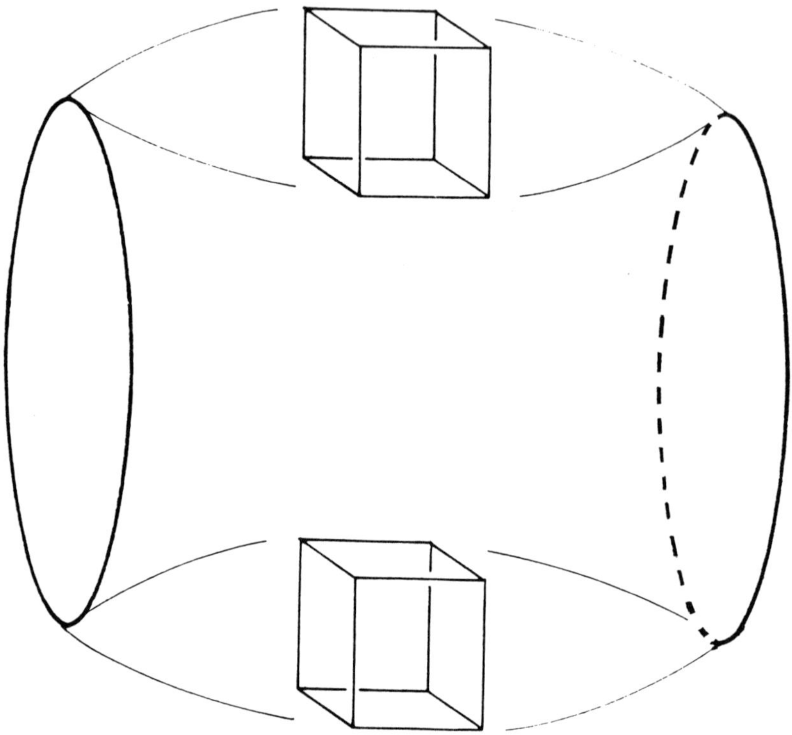

The Cartesian product $X = \Sigma T^3 \times S^1$ of the circle with the suspension of the 3-torus

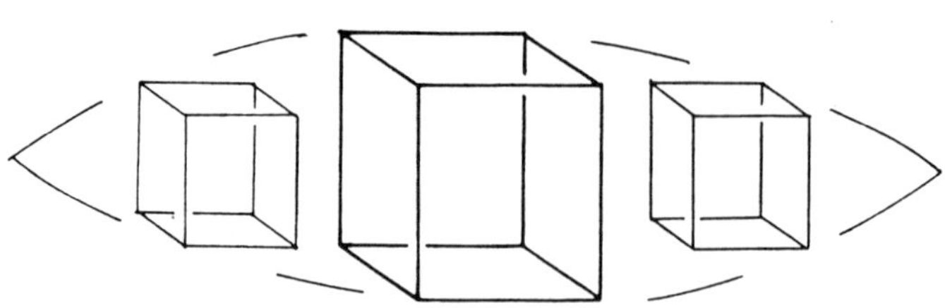

The suspension ΣT^3 of the 3-torus, showing three level sets.

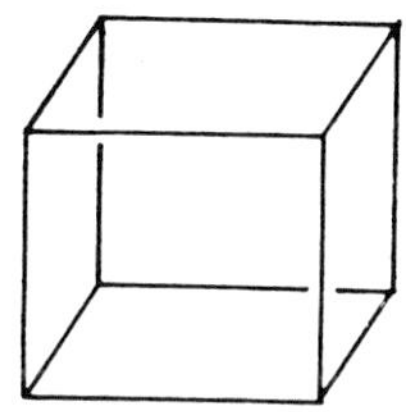

The 3-torus T^3
(identify opposite faces)

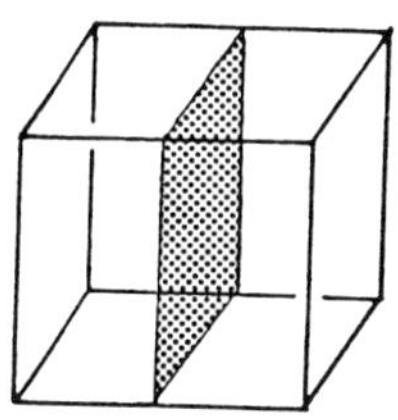

The 2-cycle T_a^2

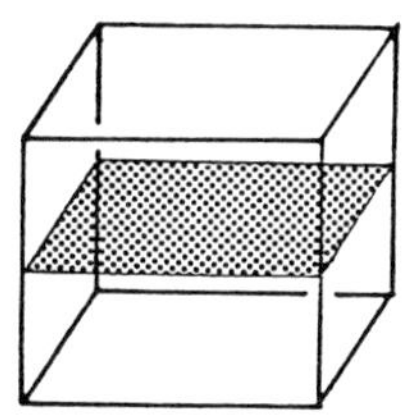

The 2-cycle T_b^2

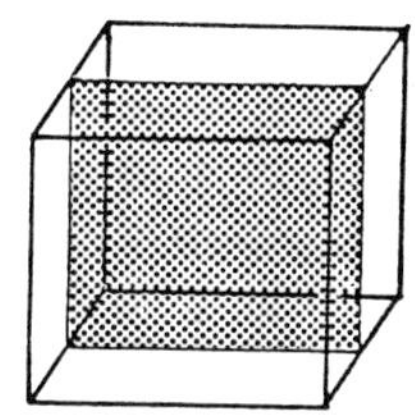

The 2-cycle T_c^2

The 1-cycle T_a^1

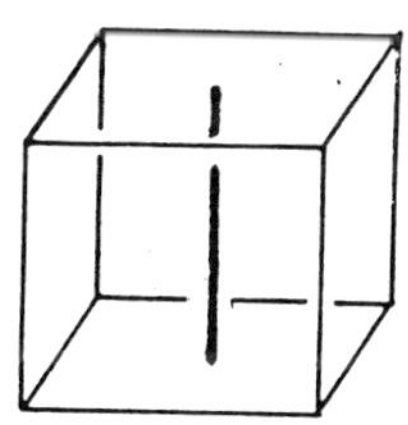

The 1-cycle T_b^1

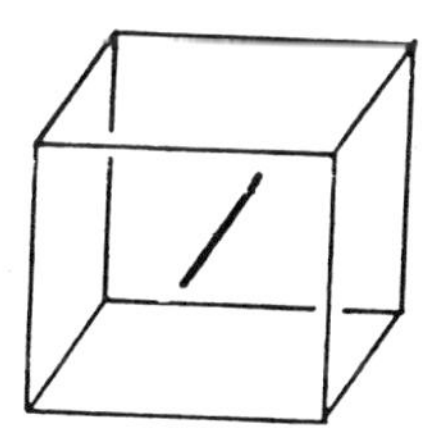

The 1-cycle T_c^1

Intersection Cohomology
Progress in Mathematics, Vol. 50
© 1984 Birkhäuser Boston, Inc.

IV STRUCTURES DE PSEUDOVARIÉTÉ SUR LES ESPACES ANALYTIQUES COMPLEXES

N. A'Campo

Le but de cet exposé est de munir un espace analytique complexe d'une structure de pseudovariété. On distinguera quatre types de pseudovariété : topologique, Lipschitz, PL et différentiable.

Un espace analytique admet une structure de pseudovariété topologique canonique (Whitney, Thom, Mather, Teissier); Dennis Sullivan m'a annoncé son travail en commun avec W. Hardt dans lequel une structure de pseudovariété Lipschitz est obtenue pour les espaces analytiques. L'existence d'une structure PL n'est pas connue; un exemple simple de Whitney montre qu'il n'existe pas de structure différentiable en général.

§ 1 *Stratifications de Whitney.*

Soit X un espace analytique purement de dimension d , soient $Y \subset X$ un sous-espace localement fermé de X et $\eta \in Y$ un point lisse de Y . Soient $(X,\eta) \subset (\mathbb{C}^N,0)$ un plongement analytique local tel que (Y,η) soit un facteur linéaire $(\mathbb{C}^k,0)$ dans $(\mathbb{C}^N,0)$ et $r : \mathbb{C}^N \to \mathbb{C}^k$ une rétraction linéaire. L'espace tangent de X en un point lisse x de X est considéré comme un point T_xX dans $Grass_d\mathbb{C}^N$. De même pour T_yY dans $Grass_k\mathbb{C}^N$. Pour un point $x \in X - Y$, assez proche de η , on note $[r(x),x]$ la direction de la droite de $\mathbb{C}^N$ passant par $x \in X$ et $r(x) \in Y$.

Voici deux propriétés d'incidence entre X et Y au point $\eta \in Y$:

Définition : (Propriété *a* de Whitney) : Pour toute suite $(x_i)_{i\in\mathbb{N}}$ de points lisses de X , telle que $\lim x_i = \eta$ et que la suite $T_{x_i}X$ converge dans $Grass_d\mathbb{C}^N$, on a l'inclusion $\lim T_{x_i}X \supset T_\eta Y$.

(Propriété *b* de Whitney) : Pour toute suite (x_i) de points lisses de

$X - Y$, telle que $\lim x_i = \eta$, la suite $T_{x_i} X$ converge dans $Grass_d \mathbb{C}^N$ et la suite des directions $[r(x_i), x_i]$ converge dans $\mathbb{P}^1(\mathbb{C}^N)$, on a $\lim T_{x_i} X \supset \lim [r(x_i), x_i]$.

Dans l'exemple suivant on a la propriété *a* mais pas la propriété *b*:

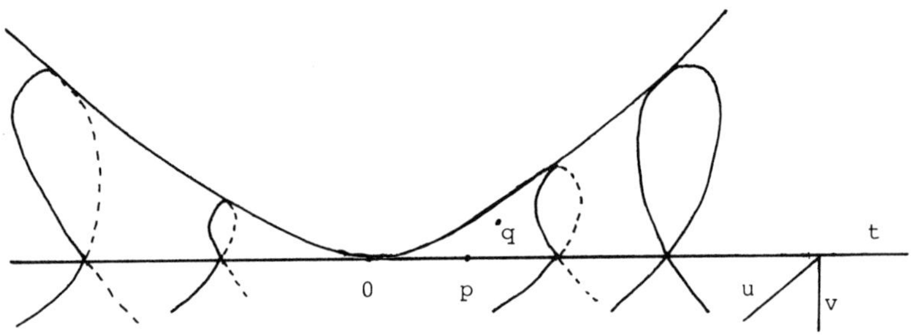

Ici $(X,0) \subset (\mathbb{C}^3,0)$, $\dim X = 2$, $\dim Y = 1$, $X = \{(u,v,t) \in \mathbb{C}^3 | u^2 = v^3 + t^2u^2\}$, Y est l'axe des t .

B. Teissier a introduit une multiplicité $M_\eta(X,Y) \in \mathbb{N}^{d-1}$ telle que l'on ait:

PROPOSITION : *Soient* X *un espace analytique,* $Y \subset X$ *un sous-espace lisse de* X *et* U *un voisinage connexe, assez petit, de* η *dans* X , *tel que* $(X - Y) \cap U$ *et* $Y \cap U$ *soient lisses. Alors les assertions suivantes sont équivalentes:*

(i) $\eta \in Y \cap U \Longrightarrow M_\eta(X,Y)$ *est constante,*

(ii) *pour tout point* $\eta \in U \cap Y$ *les propriétés a et b de Whitney entre* X *et* Y *au point* η *sont vraies.*

En outre B. Teissier a montré que la multiplicité $M_\eta(X,Y)$ ne dépend que de la restriction de l'anneau 0_X à Y et que la fonction $\eta \in X \to M_\eta(X,Y)$ est constructible par sommes et différences de fonctions caractéristiques de sous-variétés analytiques.

Soit X un espace analytique complexe. Il existe une stratification $(X_a)_{a \in A}$ de X telle que les strates X_a soient des sous-variétés lisses localement fermées de X et que pour tout couple (X_a, X_b) de strates avec $X_b \subset \bar{X}_a - X_a$ on ait en tout point de X_b les propriétés d'incidence *a* et *b* de Whitney entre X_b et $\bar{X}_a$.

Voici la construction de la multiplicité $M_\eta(X,Y)$: soient

$(X,\eta) \subset (\mathbb{C}^N,0)$ on plongement et $\underline{p} = (p_1,\dots,p_{d-1})$ un drapeau assez général de projections

$$\mathbb{C}^N \xrightarrow{p_1} \mathbb{C}^d \xrightarrow{p_2} \mathbb{C}^{d-1} \longrightarrow \dots \xrightarrow{p_{d-1}} \mathbb{C}^2 .$$

Pour $1 \leq i \leq d$, notons K_i le lieu critique de la restriction de p_i à la partie lisse de X .

Définition . On appelle i-ième variété polaire de X l'adhérence dans X de K_i . On note P_i une variété polaire. La multiplicité $M_\eta(X,Y)$ est la suite $(m_\eta(P_1),m_\eta(P_2),\dots,m_\eta(P_{d-1}))$ des multiplicités des variétés polaires au point $\eta \in Y$.

Attention : Les variétés polaires (P_i) dépendent du plongement local $(X,Y,\eta) \subset (\mathbb{C}^N,\mathbb{C}^d,0)$ et du drapeau de projections $\underline{p}$. On montre que, pour un plongement local donné, la suite $M_\eta(X,Y)$ est contante sur un ouvert de Zariski dans l'espace des drapeaux de projections et que cette valeur générique de $M_\eta(X,Y)$ ne dépend pas du plongement local.

Exemple: (Voir la figure)

$$M_0(X,Y) = (2,1), M_p(X,Y) = (2,0), M_q(X,Y) = (1,0).$$

On peut stratifier X par

$$X_0 = \{0\} \ , \ X_1 = Y - X_0 \ , \ X_2 = X - Y \ .$$

Cette stratification a les propriétés d'incidence *a* et *b* de Whitney et toute autre stratification de X avec les propriétés *a* et *b* est plus fine.

§ 2 *La structure de pseudovariété topologique* .

Soient X un espace analytique complexe et $(X_a)_{a\in A}$ une stratification de Whitney par des sous-variétés complexes. Deux strates X_a et X_b tels les que $X_b \subset \bar{X}_a - X_b$, ont donc les propriétés d'incidence *a* et *b* de Whitney en tout point de X_b.

Soient $\eta \in X$ et $Y = X_b$ la strate de X contenant η . Soient $(X,Y,\eta) \subset (\mathbb{C}^N,\mathbb{C}^d,0)$ un plongement local, $r : \mathbb{C}^n \to \mathbb{C}^k$ une rétraction linéaire, $k = \dim Y$, et *dist* une métrique hermitienne sur $\mathbb{C}^N$.

On déduit de la propriété *a* :

(1) il existe des nombres réels $\varepsilon > 0$ et $\delta > 0$ tels que pour tout $y \in Y$ avec $dist(\eta,y) < \delta$, la fibre $r^{-1}(y)$ rencontre toute trace de strate sur $T_\varepsilon(Y) = \{x \in X | dist(x,Y)\}$ transversalement,

et de la propriété *b* :

(2) il existe $\varepsilon > 0$ et $\delta > 0$ tels que l'on a (1) et que pour tout $y \in Y$ avec $dist(\eta,y) < \delta$ la fibre $r^{-1}(y)$ rencontre transversalement toute trace de strate sur $\partial T_{\varepsilon'}(Y) = \{x \in X | dist(x,Y) = \varepsilon'\}$ pour tout ε', $0 < \varepsilon' \leq \varepsilon$.

Fixons $\varepsilon > 0$ et $\delta > 0$ tels que l'on ait (1) et (2). Soient

$$K = \{x \in X \mid dist(x,Y) \leq \varepsilon \text{ et } dist(r(x),\eta) < \delta\}$$

$$r_K : K \to B^Y(\eta,\delta) = \{y \in Y \mid dist(y,\eta) < \delta\}.$$

Soit (K_s) la stratification de K la moins fine telle que les traces des strates X_a sur K et sur $K \cap \partial T_\varepsilon(Y)$ soient les strates K_s. Alors la stratification (K_s) de K est de Whitney et les fibres de r_K sont transverses aux strates K_s. Les restrictions de r_K aux strates K_s sont des submersions. Nous pouvons appliquer le théorème de Thom-Mather. Il s'ensuit que l'application r_K est localement triviale: plus précisément, soit $\delta > 0$ assez petit pour que $B^Y(\eta,\delta)$ soit contractile; alors il existe un homéomorphisme

$$H : B^Y(\eta,\delta) \times r_K^{-1}(\eta) \to K$$

tel que:

1) $r_K \circ H$ est la projection sur $B^Y(\eta,\delta)$,

2) pour toute strate K_s de K on a

$$H^{-1}(K_s) = B^Y(\eta,\delta) \times (K_s \cap r_K^{-1}(\eta)) .$$

Ainsi K est un voisinage "produit-stratifié" de η dans X, cette trivialisation locale n'est qu'un homéomorphisme.

Jusqu'à maintenant nous n'avons utilisé de (2) que la transversalité des traces des strates sur $\partial T_\varepsilon(Y)$ et sur $r^{-1}(\eta)$. De la transversalité des traces des strates sur $\partial T_{\varepsilon'}(Y)$, $0 < \varepsilon' \leq \varepsilon$, on déduit que

$r_K^{-1}(y) = r^{-1}(y) \cap K$ est au sens stratifié un cône: plus précisément

$$r^{-1}(y) = \text{cône}(\partial T_\varepsilon(Y) \cap X \cap r^{-1}(y))$$

où $\partial T_\varepsilon(Y) \cap X \cap r^{-1}(y) = L$ est le "link" de la strate $X_b = Y$ dans X . Le link est stratifié par (L_a) où $L_a = X_a \cap L$.

Nous avons ainsi décrit une structure de pseudovariété topologique sur l'espace analytique complexe X , ayant une stratification où les strates sont des sous-variétés complexes lisses de X . Les codimensions sont paires.

Références

J. Mather, Stratifications and mappings in Dynamical Systems, Academic Press, 1973.

B. Teissier, Variétés polaires II: multiplicités polaires, sections planes et conditions de Whitney, Algebraic Geometry (Proceedings, La Ràbida 1981), Springer L.N. 961(1982), 314-491.

R. Thom, Ensembles et morphismes stratifiés, Bull.A.M.S. 15(1969), 240-284.

H. Whitney, Tangents to analytic varieties, Ann. of Math.81(1965), 456-549.

Intersection Cohomology
Progress in Mathematics, Vol. 50
© 1984 Birkhäuser Boston, Inc.

V: SHEAF THEORETIC INTERSECTION COHOMOLOGY

A. Borel
(with the collaboration of N. Spaltenstein)

§ 1 SHEAF THEORY

As already pointed out in the introduction, some familiarity with sheaf theory, as developed in Godement [5] for instance, is assumed. This section is meant mainly to fix some notation and add some complements to [5].

We fix once and for all a commutative noetherian ring R. All sheaves on a space X are sheaves of R-modules. In this chapter, unless otherwise stated, R has finite cohomological dimension d, and all spaces are locally compact, locally completely paracompact (see 1.17), of finite cohomological dimension over R (see 1.15). We note however that most of the facts recalled in A are valid on arbitrary spaces (cf[5]).

A. *Generalities*

1.1 *A differential graded sheaf* (DGS or complex of sheaves) $\mathcal{S}^\bullet$ on X is a collection of sheaves $\mathcal{S}^i$ $(i \in \mathbb{Z})$ together with morphisms $d_i : \mathcal{S}^i \to \mathcal{S}^{i+1}$ such that $d_{i+1} \circ d_i = 0$ $(i \in \mathbb{Z})$. It is bounded (resp. above, resp. below) if $\mathcal{S}^i = 0$ for $|i| > N$ (resp. $i > N$, resp. $i < -N$) for some $N \in \mathbb{N}$. The derived sheaf $\mathcal{H}^\bullet\mathcal{S}^\bullet$ of $\mathcal{S}^\bullet$ is the DGS associated to the presheaf $U \mapsto H^\bullet(\mathcal{S}^\bullet(U))$ with zero differential. Its stalk at $x \in X$ is $H^\bullet(\mathcal{S}^\bullet_x)$.

A single sheaf $\mathcal{A}$ will often be identified to the complex $\mathcal{S}^\bullet$, defined by $\mathcal{S}^o = \mathcal{A}$ and $\mathcal{S}^i = 0$ for $i \neq 0$ and necessarily zero differential. A single degree complex is a DGS which is non-zero in at most one dimension.

We let Sh(X) be the category of sheaves on X and DGS(X) the category whose objects are the DGS on X and whose maps are chain maps of complexes of sheaves.

1.2 A morphism $f : A^\bullet \to B^\bullet$ of DGS induces a morphism $H^\bullet(f)$ of the derived sheaves. It is said to be a quasi-isomorphism (q.i.) if $H^\bullet(f)$ is an isomorphism. Two DGS $A^\bullet$, $B^\bullet$ are q.i. if there exist $C^\bullet$ and q.i. $A^\bullet \leftarrow C^\bullet \to B^\bullet$. We shall see later that this is an equivalence relation.

Much of what follows should formally be viewed as taking place in a derived category (whose objects are bounded below DGS, and where the q.i. are isomorphisms). We shall introduce it later (§ 5). Meanwhile, we agree however that an equal sign between two complexes of sheaves means a q.i. (i.e. is an isomorphism in the derived category).

1.3 A (right) *resolution* of a sheaf A on X is a q.i. $A \to B^\bullet$ where $B^\bullet$ is a DGS which is zero in strictly negative degrees. It is injective (flabby, fine, soft) if the B^i's are so. Under our assumptions A always has a c-soft (resp. injective) resolution which is zero in degrees $> \dim_R X$ (resp. $> \dim_R X + d + 1$) (cf section B). Similarly, a (right) resolution of a DGS $S^\bullet$ is a q.i. $S^\bullet \to B^\bullet$; it is injective (flabby, fine, soft) if the B^i's are so. If $S^\bullet$ is bounded (resp. bounded below), then it has an injective resolution which is bounded (resp. bounded below) (see 1.18).

1.4 *Hypercohomology*. Let $A^\bullet$ be a DGS on X and $A^\bullet \to J^\bullet$ an injective resolution. The hypercohomology $\mathbb{H}^\bullet(X;A^\bullet)$ is by definition $H^\bullet(\Gamma(J^\bullet))$. More generally, if Φ is a family of supports, then $\mathbb{H}^\bullet_\Phi(X;A^\bullet) = H^\bullet(\Gamma_\Phi(J^\bullet))$ If A is a single degree complex non-zero in degree k, this is just the space $H^\bullet_\Phi(X;A)$ of cohomology of X with coefficients in A and supports in Φ , shifted by k, i.e.:

$$\mathbb{H}^i_\Phi(X;A^\bullet) = H^{i-k}_\Phi(X;A^k) \qquad (i \in \mathbb{Z}) .$$

Assume $A^\bullet$ to be bounded below. According to the main theorem of sheaf theory [5:4.6.1], there is a spectral sequence which abuts to $\mathbb{H}^\bullet_\Phi(X;A^\bullet)$ and in which

$$E_2^{p,q} = H^p_\Phi(X;H^qA^\bullet) .$$

As a consequence a q.i. : $\mathcal{A}^\bullet \to \mathcal{B}^\bullet$ induces an isomorphism of hypercohomology.

Remark : We may also use resolutions by flabby or, if X is paracompact, by fine or soft sheaves. In particular if $\mathcal{A}^\bullet$ consists of such sheaves, we may take $\mathcal{A}^\bullet = \mathcal{J}^\bullet$.

1.5 *Derived functors*. Let T be a functor from Sh(X) to some abelian category. Then, for $\mathcal{S}^\bullet \in$ DGS(X), the derived functor $RT(\mathcal{S}^\bullet)$ is by definition $T(\mathcal{J}^\bullet)$, where $\mathcal{J}^\bullet$ is an injective resolution of $\mathcal{S}^\bullet$. The ith derived functor is then $R^iT(\mathcal{S}^\bullet) = H^i(T(\mathcal{J}^\bullet))$. In order to compute it, any T-acyclic resolution of $\mathcal{S}^\bullet$ may be used.

1.6 *Inverse and direct images*. To a continuous map $f : X \to Y$ there are associated a direct image functor $f_* : Sh(X) \to Sh(Y)$ and inverse image functor $f^* : Sh(Y) \to Sh(X)$. They satisfy the adjunction formula

$$\text{(1)} \qquad \mathrm{Hom}(f^*\mathcal{B},\mathcal{A}) = \mathrm{Hom}(\mathcal{B},f_*\mathcal{A}) \qquad (\mathcal{A} \in Sh(X),\ \ \mathcal{B} \in Sh(Y)).$$

There are in particular natural morphisms

$$\text{(2)} \qquad \mathcal{B} \to f_*f^*\mathcal{B},\ f^*f_*\mathcal{A} \to \mathcal{A} \qquad (\mathcal{A} \in Sh(X),\ \ \mathcal{B} \in Sh(Y)).$$

The functor f^* is exact, therefore f_* transforms injective sheaves into injective sheaves. It is also left exact. We recall that $f_*\mathcal{A}(V) = \mathcal{A}(f^{-1}(V))$ (V open in Y). If we view $\mathcal{B} \in Sh(Y)$ as an étale space on Y, then $f^*\mathcal{B}$ is the pull back on X via f . In particular $f^*\mathcal{B}_x = \mathcal{B}_{f(x)}$. These functors extend to the DGS. If $\mathcal{S}^\bullet \in$ DGS(X), then $R^if_*\mathcal{S}^\bullet$ is the sheaf associated to the presheaf:

$$V \longmapsto \mathbb{H}^i(f^{-1}V;\mathcal{S}^\bullet) \qquad (V \text{ open in } Y;\quad i \in \mathbb{N})$$

If f is an inclusion, then f^* is the restriction. If f is a closed inclusion, then f_* is the extension by zero.

We recall also that if $f : X \to Y$ is a continuous map and $\mathcal{S}^\bullet \in$ DGS(X), then we have a natural isomorphism

$$\text{(3)} \qquad \mathbb{H}^\bullet(X;\mathcal{S}^\bullet) = \mathbb{H}^\bullet(Y;Rf_*\mathcal{S}^\bullet) .$$

If, moreover, f is *proper* then

(4) $$\mathbb{H}^{\bullet}_c(X;\mathcal{S}^{\bullet}) = \mathbb{H}^{\bullet}_c(Y;Rf_*\mathcal{S}^{\bullet}) .$$

In fact, if $\mathcal{S}^{\bullet} \to \mathcal{J}^{\bullet}$ is an injective resolution of $\mathcal{S}^{\bullet}$, then $f_*\mathcal{J}^{\bullet}$ is an injective complex, therefore both sides of (3) (resp. (4)) are equal to $H^{\bullet}(\Gamma(\mathcal{J}^{\bullet}))$ (resp. $H^{\bullet}(\Gamma_c(\mathcal{J}^{\bullet}))$. The hypercohomology spectral sequence for the right hand side of (3) (resp. (4)) is then the Leray spectral sequence of f for closed (resp. compact) supports.

1.7 *The attachment map.* Let Z be a closed subspace of X and i the inclusion of U = X - Z into X. For $\mathcal{S}^{\bullet} \in DGS(X)$, the composition of the natural morphisms

(1) $$\mathcal{S}^{\bullet} \to i_*i^*\mathcal{S}^{\bullet} \to Ri_*i^*\mathcal{S}^{\bullet} ,$$

is the attachment map. It is of course a q.i. at every point $x \in U$. To say that it is a q.i. at $x \in Z$ in some dimension i amounts to the condition

(2) $$H^i(\mathcal{S}^{\bullet}_x) = \varinjlim \mathbb{H}^i(V - (V \cap Z);\mathcal{S}^{\bullet}) ,$$

where V runs through a fundamental set of neighborhoods of x in X.

The derived map of the attachment map is in fact part of long exact sequence involving cohomology with supports in Z, as will be described below.

1.8 *The functors* $j^!$ *and* $j_!$ *for a closed immersion.* Let Z, U be as above and $j : Z \to X$ the inclusion. For $\mathcal{A} \in Sh(Z)$ we let $j_!\mathcal{A} = j_*\mathcal{A}$ be the extension of $\mathcal{A}$ by zero. For $\mathcal{B} \in Sh(X)$, let $\gamma_Z\mathcal{B}$ be defined by

(1) $$\gamma_Z\mathcal{B}(V \cap Z) = \Gamma_{Z \cap V}(\mathcal{B};V) , \quad (V \text{ open in } X) .$$

There is again an (obvious) adjunction formula :

(2) $$\mathrm{Hom}(\mathcal{A},\gamma_Z\mathcal{B}) = \mathrm{Hom}(j_!\mathcal{A},\mathcal{B}) , \quad (\mathcal{A} \in Sh(Z),\ \mathcal{B} \in Sh(X)) .$$

Since $j_!$ is clearly exact, γ_Z transforms injective sheaves into injective sheaves.

For $\mathcal{S}^{\bullet} \in \mathrm{DGS}(X)$, set

(3) $$j^{!}\mathcal{S}^{\bullet} = R\gamma_Z\mathcal{S}^{\bullet} = \gamma_Z\mathcal{J}^{\bullet} \;, \quad (\mathcal{S}^{\bullet} \to \mathcal{J}^{\bullet} \text{ injective resolution}) \;.$$

By definition, the hypercohomology space of X with support in Z, with respect to $\mathcal{S}^{\bullet}$, is

(4) $$\mathbb{H}_Z^{\bullet}(X;\mathcal{S}^{\bullet}) = \mathbb{H}^{\bullet}(X;j_!j^{!}\mathcal{S}^{\bullet}) = H^{\bullet}(\gamma_Z\mathcal{J}^{\bullet}(Z)) \;.$$

Since $j^{!}\mathcal{S}^{\bullet}$ consists of injective sheaves on Z, it may be viewed as its own injective resolution, therefore

(5) $$\mathbb{H}_Z^{\bullet}(X;\mathcal{S}^{\bullet}) = \mathbb{H}^{\bullet}(Z;j^{!}\mathcal{S}^{\bullet}) \;.$$

Since $\mathcal{J}^{\bullet}$ is injective, (flabby would also do), the sequence

(6) $$0 \to j_!\gamma_Z\mathcal{J}^{\bullet} \to \mathcal{J}^{\bullet} \to i_*i^*\mathcal{J}^{\bullet} \to 0$$

is exact. The long exact sequence associated to it is then

(7) $$\cdots \to H^i((j^{!}\mathcal{S}^{\bullet})_x) \to H^i(\mathcal{S}^{\bullet}_x) \xrightarrow{\alpha_i} H^i((Ri_*i^*\mathcal{S}^{\bullet})_x) \to \cdots$$

where α_i is the attachment map. Globally (6) yields the long exact sequence

(0) $$\cdots \to \mathbb{H}_Z^i(X;\mathcal{S}^{\bullet}) \to \mathbb{H}^i(X;\mathcal{S}^{\bullet}) \to \mathbb{H}^i(U;\mathcal{S}^{\bullet}) \to \cdots$$

1.9 LEMMA *Let* Z' *be a closed subspace of* Z *and* $g : Z' \to Z$ *the inclusion map. Then* $(j \circ g)^{!} = g^{!} \circ j^{!}$.

It is clear from the definition that

(1) $$\gamma_{Z'}\mathcal{S}^{\bullet} = \gamma_{Z'}(\gamma_Z\mathcal{S}^{\bullet}).$$

By definition :

$$(j \circ g)^{!}\mathcal{S}^{\bullet} = \gamma_{Z'}\mathcal{J}^{\bullet},$$

where $J^\bullet$ is an injective resolution of $S^\bullet$. But

$$\gamma_{Z'}(J^\bullet) = \gamma_{Z'}(\gamma_Z J^\bullet) = \gamma_{Z'}(R\gamma_Z S^\bullet) = \gamma_{Z'}(j^! S^\bullet) ,$$

Since $j^! S^\bullet$ consists of injective sheaves, we have

$$\gamma_{Z'}(j^! S^\bullet) = R\gamma_{Z'}(j^! S^\bullet) = g^!(j^! S^\bullet) ,$$

which proves the lemma.

1.10 *Truncation.* Let $S^\bullet \in DGS(X)$ and $k \in \mathbb{Z}$. The truncation $\tau_{\leqq k} S^\bullet$ of $S^\bullet$ up to k is defined by

$$(\tau_{\leqq k} S^\bullet)^i = \begin{cases} S^i , & i < k \\ \ker d_i , & i = k \\ 0 , & i > k \end{cases}$$

Then $H^i(\tau_{\leqq k} S^\bullet) = 0$ for $i > k$ and the natural inclusion $\tau_{\leqq k} S^\bullet \to S^\bullet$ as a q.i. up to k.

Similarly, one defines $\tau^{\geqq k} S^\bullet$ by

$$(\tau^{\geqq k} S^\bullet) = \begin{cases} 0 , & \text{if } i < k \\ \operatorname{Coker} d_{i-1} , & \text{if } i = k \\ S^i & \text{if } i > k \end{cases}$$

We have $H^i(\tau^{\geqq k} S^\bullet) = 0$ for $i < k$ and the natural (surjective) morphism $S^\bullet \to \tau^{\geqq k} S^\bullet$ is a q.i. in degrees $\geqq k$.

1.11 (a) The constant sheaf on X with stalk R is denoted R_X . Given $U \subset X$ open and $S \in Sh(X)$, we let S_U denote either the restriction of S to U or the element of Sh(X) obtained by extending S_U by 0 on X - U . In particular, R_U is either the constant sheaf with stalk R on U, or the extension of the latter by 0 .

Given $S \in Sh(X)$ and a section s of S on U, there is an obvious map $R_U \to S$ which, for $x \in U$, assigns s(x) to $1 \in R_{U,x}$. From this it follows easily that S is a quotient of a direct sum of sheaves R_U [5:II,2.9.4].

(b) We recall that if the space X is contractible and A is a constant sheaf on X, then $H^i(X;A) = 0$ for $i \geqq 1$. In view of the difficulty we have to locate an easily accessible reference, we sketch a way to this theorem:

(a) Prove the theorem if $X = I$ is an interval.

(b) Let T be compact, Hausdorff, acyclic for cohomology with constant coefficients, Y a space, $\pi : Y \times T \to Y$ the natural projection and $\mathcal{B}$ a sheaf on Y. Then $\pi^\bullet : H^\bullet(Y;\mathcal{B}) \to H^\bullet(Y \times T;\pi^*\mathcal{B})$ is an isomorphism (Vietoris-Begle). To see this one considers the spectral sequence of π. Since π is proper, the fibre of the Leray sheaf $R^\bullet\pi_*\pi^*\mathcal{B}$ at $y \in Y$ is $H^\bullet(f^{-1}_{(y)}; \mathcal{B}_y)$, [4:IV,4.2]. Therefore the spectral sequence of π degenerates and yields our statement.

(c) For $t \in T$, let $i_t : Y \to Y \times T$ be defined by $y \longmapsto (y,t)$. Then $i_t^\bullet$ is an inverse to $\pi^\bullet$, hence is *independent* of t.

(d) Take now $T = I$. If X is contractible and $\sigma : X \times I \to X$ describes a homotopy of the identity to the constant map, then consider $i_t^\bullet \circ \sigma^\bullet$ and use (c).

B. *Cohomological dimension and bounded resolutions.*

In this section, for the convenience of the reader, we review in more detail and prove some basic known facts on cohomological dimension and the existence of bounded resolutions, in the form most suitable to our needs. For variants, generalizations and further details we refer to [4;5;8;10].

1.12 LEMMA *Let* Y *be a topological space and* A *a subsheaf of* R_Y.

(i) *For every ideal* $\mathfrak{m}$ *of* R, *the set* $\{y \in Y | \mathfrak{m} \subset A_y\}$ *is open in* Y.

(ii) *Let* $V = \{y \in Y | A_y = R\}$. *If* $V \neq X$, *then there exists a proper ideal* $\mathfrak{m}$ *of* R *and an open set* $U \supsetneq V$ *such that* $A_U = \mathfrak{m}_U + R_V$.

Proof : (i) follows from the fact that $\mathfrak{m}$ is finitely generated, since R is noetherian. To prove (ii), it suffices to take for $\mathfrak{m}$ a maximal element among the $A_x, x \notin V$.

1.13 LEMMA . *Let* Y *be any topological space and* $S \in \mathrm{Sh}(Y)$. *Then the following conditions are equivalent :*

(i) S *is injective.*

(ii) *For any pair of open sets* $U \supset V$, *the restriction map* $\mathcal{S}(U) \to \mathcal{S}(V)$ *is a split surjection of injective* R*-modules.*

[This lemma and its use in 1.17 were pointed out to me by N. Spaltenstein].

Proof : By a standard construction [5:II,7.1] any sheaf $\mathcal{A}$ can be embedded in an injective sheaf $\mathcal{J}$ such that, for any $U \subset Y$ open, we have $\mathcal{J}(U) = \Pi_{y \in U} \mathcal{I}(y)_y$, where $\mathcal{I}(y)$ is a skyscraper sheaf with support in $\{y\}$.The sheaf $\mathcal{J}$ clearly satisfies (ii). If now $\mathcal{A}$ is injective, then it is a direct factor of $\mathcal{J}$, hence (ii) also holds for $\mathcal{A}$.

Assume now that $\mathcal{S}$ satisfies (ii). Let $\mathcal{A} \subset \mathcal{B}$ be sheaves on Y and $f : \mathcal{A} \to \mathcal{S}$ a morphism. We have to show that f extends to $\mathcal{B}$. The set of subsheaves $\mathcal{C}$ of $\mathcal{B}$ containing $\mathcal{A}$, to which f extends, ordered by inclusion, is obviously inductive, hence, by Zorn lemma, has a maximal element. We are therefore reduced to proving that if $\mathcal{A} \neq \mathcal{B}$, then f extends to a subsheaf $\mathcal{C}$ of $\mathcal{B}$ containing $\mathcal{A}$ strictly.
Let first $\mathcal{B} = R_Y$. Since $\mathcal{A} \neq \mathcal{B}$, we may choose U,V and $\mathfrak{m}$ as in 1.12 (ii). Let $\mathcal{C} = \mathcal{A} + R_U$. By construction $\mathcal{A}_U \neq R_U$, hence $\mathcal{C}$ contains $\mathcal{A}$ strictly. We want to prove that f extends to $\mathcal{C}$. For this it suffices to extend $f|\mathcal{A}_U$ to R_U. This amounts to finding a map $f' : R \to \mathcal{A}(U)$ which makes the diagram

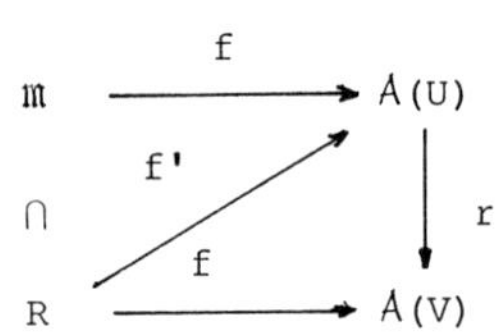

commutative. But the existence of f' follows from the fact that r is a split surjection of injective R-modules.

Since $R_U \subset R_Y$ for any open $U \subset Y$, this also proves our assertion when $\mathcal{B} = R_U$. In general, since $\mathcal{A} \neq \mathcal{B}$, we may find an open set $U \subset Y$ and a section s of $\mathcal{B}_U$ not belonging to $\mathcal{A}_U$. Use s to construct a morphism $g : R_U \to \mathcal{B}$ as recalled in 1.12. Then

$$\mathcal{A} \subsetneq \mathcal{C} = \mathcal{A} + g(R_U) \subset \mathcal{B}$$

and the above readily yields an extension of f to $\mathcal{C}$.

1.14 Let Φ be a family of supports in X and

(1) $$0 \to A \to J^o \to J^1 \to \dots \to J^{n-1} \to B \to 0$$

an exact sequence in Sh(X) where the J^i's are acyclic in Φ- cohomology. Then, by using long exact sequences in cohomology, one gets

(2) $$H^i_\Phi(U;B) = H^{i+n}_\Phi(U;A) \qquad (U \subset X \text{ open } ; \ i \geqq 1)$$

In particular, taking for Φ the family of compact subsets, we see that the condition

(3) $$H^{n+1}_c(U;A) = 0 \qquad \text{for all U open in X}$$

implies $H^1_c(U;B) = 0$ for all such U's, i.e. that B is c-soft. Since A always has a c-soft resolution $K^\bullet$, this shows that if A satisfies (3) then it has a c-soft resolution which vanishes in degrees > n, (namely $\tau_{\leqq n}K^\bullet$) .

1.15 DEFINITION . The cohomological dimension $\dim_R X$ of X over R is the smallest $n \in \mathbb{N} \cup \infty$ such that

(1) $$H^i_c(U;A) = 0 \text{ for all U open in X}, A \in Sh(Y) \text{ and } i > n.$$

Note that, by 1.14, the condition (1) for i = n + 1 implies (1) in general.

1.16 PROPOSITION . *Let* X *be a locally compact space and* $n \in \mathbb{N}$. *Then the following conditions are equivalent*

(i) $\dim_R X \leqq n$.

(ii) $H^{n+1}_c(X;A) = 0$ *for all* $A \in Sh(X)$.

(iii) $H^{n+1}_c(U;\mathfrak{m}) = 0$ *for every ideal* $\mathfrak{m}$ *of* R *and every open subset* U *of* X.

(iv) $H^{n+1}_c(U;R) = 0$ *for all* U *open in* X.

Proof : The implications (i) $\Rightarrow$ (ii) and (iii) $\Rightarrow$(iv) are obvious. The equalities

(1) $$H^\bullet_c(U;\mathcal{B}_U) = H^\bullet_c(U;\mathcal{B}) = H^\bullet_c(X;\mathcal{B}_U), \quad (U \text{ open in } X;\ \mathcal{B} \in Sh(X)$$

show that (ii) $\Rightarrow$ (iii) and, taking 1.14 into account, that (ii) $\Rightarrow$ (i). There remains to prove that (iv) $\Rightarrow$ (iii) $\Rightarrow$ (ii).

Assume (iv) to hold. Then, by 1.14, R_U has a c-soft resolution vanishing in degrees > n . Therefore

(2) $$H^i_c(U;R) = 0 \text{ for all } i > n, \quad (U \text{ open in } X).$$

The same is then true with R replaced by $R^m (m \in \mathbf{N})$ or also by any finitely generated projective R-module P, since such a module is a direct summand of an R^m. Since R has finite dimension d , any $\mathfrak{m}$ has a left resolution of length $\leq$ d by finitely generated projective R-modules. Then (iii) follows from the above and 1.14.

Assume now (iii) to hold. Let M be the family of all sheaves $\mathcal{S}$ for which $H^{n+1}_c(X;\mathcal{S}) = 0$. We have to show that M = Sh(X).

By assumption, M contains the sheaves $\mathfrak{m}_U$ ($\mathfrak{m}$ ideal of R;U open in X). It is also clear that if $0 \to \mathcal{A}' \to \mathcal{A} \to \mathcal{A}'' \to 0$ is an exact sequence in Sh(X) and $\mathcal{A}', \mathcal{A}'' \in M$ (resp. $\mathcal{A}',\mathcal{A} \in M$) then $\mathcal{A} \in M$ (resp. $\mathcal{A}'' \in M$). Finally, if $(A_i)_{i\in I}$ is directed under inclusion and consists of elements in M, then the union of the A_i belongs to M since cohomology with compact supports commutes with inductive limits [5:II,4.12.1]. But it is known that this implies M = Sh(X), (see [4:15.10]e.g).

1.17 PROPOSITION . *Assume that* $\dim_R X$ *is finite and that* X *is locally completely paracompact. Let* $n = \dim_R X$ *and* d = dim R . *Let*

(1) $$0 \to \mathcal{A} \to \mathcal{J}^0 \xrightarrow{d_0} \mathcal{J}^1 \xrightarrow{d_1} \cdots \to \mathcal{J}^{m-1} \xrightarrow{d_{m-1}} \mathcal{B} \to 0$$

be an exact sequence in Sh(X).

(i) *If the* $\mathcal{J}^i$*'s are flabby and* m = n + 1, *then* $\mathcal{B}$ *is flabby.*

(ii) *If the* $\mathcal{J}^i$*'s are injective and* m = n + d + 1, *then* $\mathcal{B}$ *is injective.*

[We say that a space is locally completely paracompact if every point has an open neighborhood all of whose open subsets are paracompact.]

Proof : (i) Let $Z^n = \ker d_n$. By 1.14, Z^n is c-soft. We have the short exact sequence

$$(2) \qquad 0 \to Z^n \to J^n \to \mathcal{B} \to 0 .$$

Let U be an open paracompact subspace of X. By Theorem 3.4.1 in [5:II], the restriction of Z^n to U is soft, hence

$$(3) \qquad H^i(U;Z^n) = 0 , \qquad (i \geqq 1) .$$

The exact sequence (1) then gives rise to the commutative diagram

$$(4) \qquad \begin{array}{ccccc} J^n(X) & \longrightarrow & \mathcal{B}(X) & & \\ \downarrow & & \downarrow & & \\ J^n(U) & \longrightarrow & \mathcal{B}(U) & \longrightarrow & 0 \\ \downarrow & & & & \\ 0 & & & & \end{array}$$

with exact first column since J^n is flabby and exact second row by (3). Therefore $\mathcal{B}(X) \to \mathcal{B}(U)$ is surjective. X being assumed to be locally completely paracompact, it follows that every point has an open neighborhood on which the restriction of $\mathcal{B}$ is flabby. Since flabbiness is a local property [5:II,3.11], this proves (i).

(ii) Let $Z^j = \ker d_j (n < j \leqq n + d)$. By (i) Z^{n+1} is flabby. Since the J^i's are injective, hence a fortiori flabby, the equalities

$$(5) \qquad Z^{n+i+1} = J^{n+i}/Z^{n+i} \quad (1 \leqq i < d), \ \mathcal{B} = J^{n+d}/Z^{n+d}$$

imply that Z^{n+i} $(1 \leqq i \leqq d)$ and $\mathcal{B}$ are also flabby. In particular, all terms in the exact sequence

$$(6) \qquad 0 \to Z^{n+1} \to J^{n+1} \to \ldots \to J^{n+d} \to \mathcal{B} \to 0$$

are flabby. In order to prove that $\mathcal{B}$ is injective it suffices, by 1.13, to show that given $V \subset U$ open in X, the restriction map $r : \mathcal{B}(U) \to \mathcal{B}(V)$ is a split surjection of injective modules. Given $V \subset U$, we get from

(6) the commutative diagram

$$\begin{array}{ccccccccccc}
 & & 0 & & 0 & & 0 & & 0 & & \\
 & & \downarrow & & \downarrow & & \downarrow & & \downarrow & & \\
0 & \to & C^n & \longrightarrow & C^{n+1} & \to \cdots \to & C^{n+d} & \longrightarrow & C^{n+d+1} & \longrightarrow & 0 \\
 & & \downarrow & & \downarrow & & \downarrow & & \downarrow & & \\
0 & \to & Z^{n+1}(U) & \longrightarrow & J^{n+1}(U) & \to \cdots \to & J^{n+d}(U) & \longrightarrow & \mathcal{B}(U) & \longrightarrow & 0 \\
 & & \downarrow & & \downarrow & & \downarrow & & \downarrow & & \\
0 & \to & Z^{n+1}(V) & \longrightarrow & J^{n+1}(V) & \to \cdots \to & J^{n+d}(V) & \longrightarrow & \mathcal{B}(V) & \longrightarrow & 0 \\
 & & \downarrow & & \downarrow & & \downarrow & & \downarrow & & \\
 & & 0 & & 0 & & 0 & & 0 & &
\end{array}$$

where the C^{n+i} $(0 \leqq i \leqq d+1)$ are the kernels of the restriction maps. The columns are exact since the sheaves Z^{n+1}, J^{n+i} and $\mathcal{B}$ are flabby. This last fact also implies that the second and third rows are exact. Then so is the first row. By 1.13, $J^{n+i}(U)$, $J^{n+i}(V)$ and C^{n+i} are injective modules $(i = 1,\ldots,d)$. Since $\dim R = d$, it follows by standard homological algebra that C^{n+d+1}, $\mathcal{B}(U)$ and $\mathcal{B}(V)$ are injective modules. Then r is also a split surjection, whence our assertion.

1.18 COROLLARY. *Any* $A \in \mathrm{Sh}(X)$ *has a flabby (resp. injective) resolution* $A \to J^{\bullet}$ *which vanishes in degree* $> n+1$ *(resp.* $> n+d+1$*). Any* $A^{\bullet} \in \mathrm{DGS}(X)$ *which is bounded (resp. bounded below) has an injective resolution which is bounded (resp. bounded below).*

In fact, A always has a flabby (resp. injective) resolution $A \to J^{\bullet}$ By the theorem, $\tau_{\leqq n+1}J^{\bullet}$ (resp. $\tau_{\leqq n+d+1}J^{\bullet}$) is then a flabby (resp. injective) resolution. This proves the first assertion. The second one then follows from the construction of a resolution of $A^{\bullet}$ by means of resolutions of the A^i's. More specifically, if

$$A^m \to I^{m,0} \to I^{m,1} \to \ldots \qquad (m \in \mathbb{Z})$$

is the Godement injective resolution of A^m [5:II,7.1] then, by 1.17,

$$A^m \to J^{m,\bullet} = \tau_{\leqq e}I^{m,\bullet} \qquad (e = n+d+1)$$

is also an injective resolution, which is obviously bounded. Moreover there is a morphism $J^{m,i} \to J^{m+1,i}$ $(i \in \mathbb{N})$ which sits over $d : A^m \to A^{m+1}$. From this we get on the direct sum $J^{\bullet,\bullet}$ of the $J^{m,i}$ a structure of double complex. Let then $J^\bullet$ be the simple complex derived from $J^{\bullet,\bullet}$ by using the total degree. Then $A^\bullet \to J^\bullet$, where $A^m \to J^m$ is defined via $A^m \to J^{m,0}$, is the sought for injective resolution.

It is also possible to combine 1.17 with the construction of an injective resolution given in [8:I,7.1].

§ 2 DELIGNE'S SHEAF. FIRST AXIOMATIC CHARACTERIZATION

2.1 Let X be a Hausdorff space. A filtration $\mathfrak{X} = (X_i)$

(1) $$X = X_n \supset X_{n-2} \supset X_{n-3} \supset \dots \supset X_0 \supset X_{-1} = \emptyset$$

by closed subspaces is said to be a n-*dimensional topological stratification* if it satisfies the conditions of 1.1 in I . This implies in particular that X is locally compact, of cohomological dimension n .The space X is a n-*dimensional topological pseudomanifold* if it admits a topological stratification, a *stratified* n-*dimensional pseudomanifold* if it has been endowed with one.

In this section, X is a stratified n-dimensional pseudomanifold $\mathfrak{X}$ its stratification.

The stratum $S_i = X_i - X_{i-1}$ is an i-manifold, (if not empty). We set

(2) $$U_i = X - X_{n-i} \qquad (2 \leqq i \leqq n+1) ,$$

The U_i's form an increasing sequence of open subsets and we have

(3) $$U_{k+1} = U_k \cup S_{n-k} .$$

We let i_k and j_k be the inclusions

(4) $$U_k \xrightarrow{i_k} U_{k+1} \xleftarrow{j_k} S_{n-k} .$$

If $\mathcal{S}^\bullet$ is a DGS on X, we let $\mathcal{S}^\bullet_k$ denote its restriction to U_k . We let $\mathcal{E}$ be a local system on U_2 , i.e. a locally constant sheaf of R-modules with finitely generated stalks. If U_2 is connected, it is defined in the usual way by a finitely generated R-module E_0 over the fundamental group $\pi_1(U_2)$ of U_2 and conversely.

Remark. In [6], Goresky and MacPherson start from a slightly more general notion of topological stratification : they allow a first stratum S_{n-1} of dimension n-1 and $X - X_{n-1}$ is not necessarily dense. However, soon after a blanket assumption requires all topological stratifications to satisfy the conditions imposed above. Therefore we do not really need this more general concept.

These restrictions are necessary to prove topological invariance, but do not intervene in the discussion of constructibility or Verdier duality and biduality. In order to be able to point that out conveniently, we shall call "unrestricted" those more general topological stratifications and, for brevity, also call pseudomanifold a space admitting one. For those we then allow $i = 1$ in (2), (3), (4).

2.2 *Deligne's sheaf*. We fix a perversity p and a local system $\mathcal{E}$ on U_2. Deligne's sheaf is defined inductively on U_k by the rules :

(1) $$\mathcal{P}(\mathcal{E})^\bullet_2 = \mathcal{E} \text{ on } U_2 ,$$

and, for $k \geqq 2$, assuming $\mathcal{P}(\mathcal{E})^\bullet_k$ defined over U_k,

(2) $$\mathcal{P}(\mathcal{E})^\bullet_{k+1} = \tau_{\leqq p(k)} Ri_{k*}\mathcal{P}(\mathcal{E})^\bullet_k .$$

Then $\mathcal{P}(\mathcal{E})^\bullet = \mathcal{P}(\mathcal{E})^\bullet_{n+1}$.
This sheaf depends on $\mathfrak{X}$, p and $\mathcal{E}$. It will be denoted by $\mathcal{P}^\bullet$, $\mathcal{P}^\bullet(\mathcal{E})$, $\mathcal{P}^\bullet_p(\mathcal{E})$, $\mathcal{P}^\bullet_{p,\mathfrak{X}}(\mathcal{E})$ according to the needs of the context. Note that no PL structure is involved in its construction. In fact, the latter makes only use of the filtration 2.1(1) of X.

2.3 *The set of axioms* $AX1_{p,\mathfrak{X}}$. Let $S^\bullet \in DGS(X)$. The AX1 consists of the following three sets of conditions

(a) *$S^\bullet$ is bounded, $S^i = 0$ for $i < 0$ and $S^\bullet_2 = \mathcal{E}$.*
(b) *For $x \in S_{n-k}$ we have $H^i(S^\bullet_x) = 0$ if $i > p(k)$, $(k = 2,\dots,n)$.*
(c) *The attachment map $\alpha_k : S^\bullet_{k+1} \to Ri_{k*}S^\bullet_k$ is a q.i. up to $p(k)$.*

2.4 LEMMA. *Assume $S^\bullet$ satisfies* AX1. *Then*

(1) $$S^\bullet_{k+1} = \tau_{\leqq p(k)} Ri_{k*}S^\bullet_k .$$

Proof : Consider the diagram :

$$(2)\qquad \begin{array}{ccc} S^{\bullet}_{k+1} & \xrightarrow{\alpha_k} & Ri_{k*}S^{\bullet}_k \\ \Big\uparrow{\scriptstyle \beta_{k+1}} & & \Big\uparrow{\scriptstyle \gamma_{k+1}} \\ \tau_{\leq p(k)}S^{\bullet}_{k+1} & \xrightarrow{\alpha'_k} & \tau_{\leq p(k)}Ri_{k*}S^{\bullet}_k \end{array}$$

By (b), β_{k+1} is a q.i. By (c) α'_k is a q.i. This implies, by definition, that $S^{\bullet}_{k+1}$ and $\tau_{\leq p(k)}Ri_{k*}S^{\bullet}_k$ are q.i.

2.5 THEOREM . *The sheaf* $P(E)^{\bullet}$ *satisfies* AX1. *Any* $S^{\bullet}$ *satisfying* AX1 *is q.i. to* $P(E)^{\bullet}$.

This first assertion is clear from the definition of $P(E)^{\bullet}$. Assume $S^{\bullet}$ satisfies AX1. Then $S^{\bullet}_2 = P(E)^{\bullet}_2$ by (a). Assume that $S^{\bullet}_k = P(E)^{\bullet}_k$ for some $k \geq 2$. Then $Ri_{k*}S^{\bullet}_k = Ri_{k*}P(E)^{\bullet}_k$, hence also

$$(1)\qquad \tau_{\leq p(k)}Ri_{k*}S^{\bullet}_k = \tau_{\leq p(k)}Ri_{k*}P(E)^{\bullet}_k$$

which yields $S^{\bullet}_{k+1} = P(E)^{\bullet}_{k+1}$ by 2.4 and 2.2 (2).

2.6 *Notation* . We let $I_pH^{\bullet}(X;E)$ denote the hypercohomology of X with respect to any $S^{\bullet}$ satisfying AX1 and call it the intersection cohomology of X with coefficients in the local system E . More generally, if Φ is a family of supports, $I_pH^{\bullet}_{\Phi}(X;E)$ is the intersection cohomology of X with coefficients in E and supports in Φ .

2.7 *Remarks.* (a) The set of axioms AX1 is not exactly the same as in [6]. There it is also required that $S^{\bullet}$ be "constructible". We shall see that this last condition is in fact a consequence of AX1 as defined here (§ 3).

(b) For simplicity, we required $S^{\bullet}$ to be bounded and zero in strictly negative degrees. But we are really "in the derived category". We could instead require only

(1) $S^{\bullet}$ *is bounded below.* $H^iS^{\bullet}$ *is zero for* $i < 0$ *and* i *big enough.*

In fact,if $S^{\bullet}$ is bounded below, the hypercohomology spectral sequence converges. Moreover, using double truncation, we see that $S^{\bullet}$ is q.i. to a bounded complex $T^{\bullet}$ which is zero in strictly negative degrees (1.10). It follows then that 2.5 extends also to such complexes of sheaves.

2.8 THEOREM . *Assume* $p = 0$ *is the zero perversity. Let* $\mathcal{E}$ *be a local system on* U_2 *and* $i : U_2 \to X$ *the inclusion. Then* $i_*\mathcal{E}$ *satisfies* AX1. *In particular*

$$(1) \qquad I_0H^{\bullet}(X;R) = H^{\bullet}(X;i_*E) .$$

If X *is normal and* $\mathcal{E} = R_{U_2}$ *, then* $i_*\mathcal{E} = R_X$ *and*

$$(2) \qquad I_0H^{\bullet}(X;R) = H^{\bullet}(X;R) .$$

Proof. Obviously, $i_*\mathcal{E}$ satisfies AX1 (a), (b). For $k \geq 2$, let $i_{2,k} : U_2 \to U_k$ be the inclusion map. It is clear from the definitions that, for any sheaves $\mathcal{A} \in Sh(U_2)$ and $\mathcal{B} \in Sh(U_k)$:

$$(3) \qquad (i_*\mathcal{A})_{k+1} = i_{2,k+1*}\mathcal{A} = i_{k*}(i_{2,k*}\mathcal{A}) \text{ and } \tau_{\leq 0}Ri_{k*}\mathcal{B} = i_{k*}\mathcal{B} .$$

This shows that $i_*\mathcal{E}$ satisfies also AX1 (c). Assume now X to be normal. Then a point $x \in S_{n-k}$ has a fundamental set of distinguished neighborhoods U such that $U - \Sigma$ is connected. This implies immediately that $i_*R_{U_2}$ is the constant sheaf R_X, whence the last assertion.

2.9 PROPOSITION . *Assume* X *to be paracompact. Let* $\mathcal{O}$ *be the orientation sheaf. Let* $I_p\mathcal{C}_{\bullet}$ *be the intersection homology sheaf defined with respect to a* PL *structure (see* I) *and let* $I_p\mathcal{C}^{\bullet}$ *be defined by* $I_p\mathcal{C}^i = I_p\mathcal{C}_{n-i}$. *Then* $I_p\mathcal{C}^{\bullet}$ *satisfies* AX1 *with* $\mathcal{E} = \mathcal{O}$. *We have*

$$(1) \qquad I_pH_i(X;R) = I_pH^{n-i}(X;\mathcal{O}) , \quad (i \in \mathbb{Z})$$

and $I_pH_{\bullet}(X;R)$ *is independent of the underlying* PL *structure.*

On U_2 , the sheaf $I_p\mathcal{C}_{\bullet}$ is the homology sheaf. Since U_2 is a manifold, we have $I_p\mathcal{C}^{\bullet} = \mathcal{O}$. Hence $I_p\mathcal{C}^{\bullet}$ satisfies AX1 (a) for $\mathcal{E} = \mathcal{O}$. That it fullfills the conditions (b) and (c) was proved in II,6.1.
By definition

$$(2) \qquad I_pH_i(X;R) = H^{n-i}(\Gamma_X(I_p\mathcal{C}^{\bullet})) .$$

But $I_p C^{\cdot}$ is soft (cf II,5.1) therefore the right hand side of (2) also represents hypercohomology, hence (1).

2.10 *Distinguished neighborhoods.* We fix here some notation to be used frequently in the sequel.

Let $k \geqq 2$ and $x \in S_{n-k}$. The distinguished neighborhoods U of x are of the form $B^{n-k} \times \mathring{c}(L)$, where B^{n-k} is an open ball in S_{n-k} around x and $\mathring{c}(L)$ an open cone over a compact pseudomanifold L of dimension k-1, the "link" of x. A distinguished neighborhood of x is also one of $y \in B^{n-k}$. By a *fundamental set* $\mathfrak{U}$ *or* $\mathfrak{U}_x$ *of distinguished neighborhoods* of x we shall always mean a countable fundamental set of neighborhoods of x obtained from one such U by shrinking B^{n-k} and $\mathring{c}(L)$ to x in the standard way. We let $\pi : U \to \mathring{c}(L)$ be the natural projection.

By definition there is a topological stratification $\mathfrak{L}$ of L such that the stratification induced by $\mathfrak{X}$ on U is the π - inverse image of the cone over $\mathfrak{L}$. We have in particular an increasing sequence of open subsets V_j of L such that

$$(1) \qquad U_j \cap U = B^{n-k} \times \mathring{c}(V_j)^* \ , \quad (j = 2,\ldots,k, \mathring{c}(V_j)^* = \mathring{c}(V_j) - \{x\}) \ .$$

If $\tilde{S}_j$ is the j-dimensional stratum of $\mathfrak{L}$, then

$$(2) \qquad V_{\ell+1} = V_\ell \cup \tilde{S}_{k-\ell-1} \ , \qquad (2 \leqq \ell < k) \ .$$

We let $\bar{i}_\ell$, $\bar{j}_\ell$ be the inclusions

$$(3) \qquad V_\ell \xrightarrow{\bar{i}_\ell} V_{\ell+1} \xleftarrow{\bar{j}_\ell} \tilde{S}_{k-1-\ell} \ , \quad (2 \leqq \ell < k) \ .$$

We have the cartesian squares

$$(4) \qquad \begin{array}{ccc} U_j \cap U & \xrightarrow{i_j} & U_{j+1} \cap U \\ \downarrow \pi & & \downarrow \pi \\ \mathring{c}(V_j)^* & \xrightarrow{c(\bar{i}_j)} & \mathring{c}(V_{j+1})^* \end{array} \qquad (2 \leqq j < k) \qquad \begin{array}{ccc} U_k \cap U & \xrightarrow{i_k} & U \\ \downarrow \pi & & \downarrow \pi \\ \mathring{c}(L)^* & \longrightarrow & \mathring{c}(L) \end{array}$$

The space $\mathring{c}(L)^*$ obtained from $\mathring{c}(L)$ by deleting the vertex is naturally isomorphic to $\mathbb{R} \times L$, hence

(5) $$U - B^{n-k} = B^{n-k+1} \times L \ .$$

We let μ (resp. ν) be the projection of the left hand side onto B^{n-k+1} (resp. L). Thus ν is the composition of π and of the natural projection of $\mathbb{R} \times L$ onto L. We also have a cartesian diagram

(6) $$\begin{array}{ccc} U_j \cap U & \xrightarrow{i_j} & U_{j+1} \cap U \\ \nu \downarrow & & \downarrow \nu \\ V_j & \xrightarrow{\bar{i}_j} & V_{j+1} \end{array}$$

2.11 Let $\mathcal{C}.$ be the homology sheaf of X, (say defined via a PL structure as in I , a more general definition will be recalled in § 7). We define $\mathcal{D}^{\bullet}$ by

(7) $$\mathcal{D}^i = \mathcal{C}_{n-i} \ , \qquad (i \in \mathbb{Z}) \ .$$

2.12 THEOREM . *Assume* X *to be normal. Let* $\mathcal{D}^{\bullet}$ *be as above and* $\mathcal{O}$ *be the orientation sheaf on* U_2 . *Then* $\mathcal{D}^{\bullet}$ *satisfies* $AX1_{t,\mathcal{O}}$ *,(where* t *is the maximal perversity). In particular*

(1) $$I_tH^i(X;\mathcal{O}) = H_{n-i}(X;R) \qquad (i \in \mathbb{Z}) \ .$$

Proof. We have $\mathcal{D}^{\bullet}_2 = \mathcal{O}$ in U_2 since U_2 is a manifold, hence $\mathcal{D}^{\bullet}$ satisfies AX1 (a). We have $t(k) = k - 2$ $(k \geq 2)$. To prove that $\mathcal{D}^{\bullet}$ satisfies $AX1_t$ (b), (c), we have then to show, for $k = 2,\dots,n$:

(2) $$H^j(\mathcal{D}^{\bullet}_x) = 0 \qquad (x \in S_{n-k}, j > k - 2)$$

(3) $$\alpha_k : H^j(\mathcal{D}^{\bullet}_{k+1,x}) \to H^j((Ri_{k*}\mathcal{D}^{\bullet}_k)_x) \text{ is an isomorphism for } j \leq k - 2$$

In the sequel, homology is understood with coefficients in R. In the notation of 2.10,we have

(4) $$H^j(\mathcal{D}^{\bullet}_x) = \varinjlim_U H_{n-j}(U) \ ,$$

(5) $$H^j((Ri_{k*}\mathcal{D}^{\bullet}_k)_x) = \varinjlim_U H_{n-j}(U - B^{n-k}) \ ,$$

as U runs through a fundamental set $\mathfrak{U}$ of distinguished neighborhoods of x (2.10). It is therefore enough to prove the two following assertions

(6) $$H_i(U) = 0 \qquad (i \leqq n-k+1)$$

(7) *The restriction map* $\pi_i : H_i(U) \to H_i(U - B^{n-k})$ *is an isomorphism for* $i > n-k+1$.

We consider the long exact sequence in homology of U mod B^{n-k} :

(8) $$\ldots \to H_i(B^{n-k}) \to H_i(U) \to H_i(U - B^{n-k}) \xrightarrow{\delta_i} H_{i-1}(B^{n-k}) \to \ldots$$

Recall that $H_i(B^{n-k})$ is zero for $i \neq n-k$ and equal to R for $i = n-k$. Therefore (8) yields (7). We have $U - B^{n-k} = B^{n-k+1} \times L$, hence

(9) $$H_i(U - B^{n-k}) = H_{i-(n-k+1)}(L) , \quad (i \in \mathbb{Z}) .$$

This is zero if $i \leqq n-k$. Therefore (8) implies (6) for $i < n-k$. Moreover we see that in order to prove (6) for $i = n-k, n-k+1$, it suffices to show :

(10) $$\delta_{n-k+1} : H_{n-k+1}(U - B^{n-k}) \to H_{n-k}(B^{n-k}) \textit{ is an isomorphism} .$$

We have $U = B^{n-k} \times \overset{\circ}{c}(L)$ and $U - B^{n-k} = B^{n-k} \times (\mathbb{R} \times L)$. From that and the Künneth rule, we see that we have to prove that

(11) $$\delta_1 : H_1(\mathbb{R} \times L) \to H_0(x)$$

is an isomorphism, where δ_1 is the boundary homomorphism in the segment

(12) $$H_1(\overset{\circ}{c}(L)) \to H_1(\mathbb{R} \times L) \xrightarrow{\delta_1} H_0(x) \to H_0(\overset{\circ}{c}(L))$$

of the long exact sequence in homology of $\overset{\circ}{c}(L)$ modulo its vertex. We have $H_0(x) = R$. The space L is compact and, since X is normal, is connected, hence

(13) $$H_1(\mathbb{R} \times L) = H_0(L) = R .$$

moreover $\overset{\circ}{c}(L)$ is connected, non - compact, hence $H_0(\overset{\circ}{c}(L)) = 0$. As a consequence, δ_1 is an isomorphism.

§ 3 CONSTRUCTIBILITY

In this paragraph R is a noetherian commutative ring of finite cohomological dimension and X is a locally compact topological space of finite cohomological dimension over R. We assume moreover that every point of X has a countable fundamental system of neighborhoods.

3.1 We introduce first some notions concerning direct and inverse systems of R-modules. The index sets are always assumed to be filtered on the right (for any indices i and j there exists k with $k \geqq i$ and $k \geqq j$).

Definitions. A *direct system* of R-modules $(A_i)_{i \in I}$ is *essentially constant* if the following conditions hold:
1) For each $i \in I$, there exists $i' \geqq i$ such that

$$\text{Ker } (A_i \to A_{i'}) = \text{Ker } (A_i \to \varinjlim A_j) .$$

2) There exists $i_o \in I$ such that $A_{i_o} \to \varinjlim A_i$ is surjective.

An *inverse system* of R-modules $(A_i), (i \in I)$, is *essentially constant* if the following conditions hold:
1') For each $i \in I$, there exists $i' \geqq i$ such that

$$\text{Im } (A_{i'} \to A_i) = \text{Im } (\varprojlim A_j \to A_i)$$

2') There exists $i_o \in I$ such that $\varprojlim A_i \to A_{i_o}$ is injective.

3.2 *Remarks.* a) Let $(A_i)_{i \in I}$ be a direct system and let $\bar{A}_i = A_i / \text{Ker } (A_i \to \varinjlim A_j)$. Then $(\bar{A}_i), (i \in I)$, is a direct system in which all the maps are injective and $\varinjlim \bar{A}_i = \varinjlim A_i$. Condition (2) requires that $(\bar{A}_i)$ is a constant system if we consider only large enough indices.

In using (1') note that we may have

$$\text{Im } (\varprojlim A_j \to A_i) \neq \bigcap_{j \geqq i} \text{Im } (A_j \to A_i) .$$

However, this cannot happen if I has a countable cofinal subset.

b) Let $J \subset I$ be cofinal (i.e. for any $i \in I$ there exists $j \in J$ with $i \leqq j$). Then a direct (resp. inverse) system $(A_i)_{i \in I}$ is essentially constant if and only if $(A_i)_{i \in J}$ is so.

In particular, if I is the set of all neighborhoods (or open neighborhoods) of some point $x \in X$, ordered by $U \leqq V$ if $U \supset V$, the countability assumption implies that we need only to check (1) and (2) (resp. (1') and (2') for some suitable sequence of neighborhoods.

c) Let $(A_i)_{i \in I}$ be a direct system. If $\varinjlim A_i$ is finitely generated, then (2) holds. If for every $i \in I$ there exists $i'' > i$ such that $\mathrm{Im}(A_i \to A_{i''})$ is finitely generated, then (1) holds since R is noetherian.

Consider now an inverse system $(A_i)_{i \in I}$, and suppose that R is *artinian*. If $\varprojlim A_i$ is finitely generated, then (2') holds. If I has a countable cofinal subset and for every $i \in I$ there exists $i'' \geqq i$ such that $\mathrm{Im}(A_{i''} \to A_i)$ is finitely generated, then (1') holds.

3.3 Let $S^\bullet \in DGS^b(X)$. We introduce now various constructibility conditions.

(i) $S^\bullet$ *is cohomologically locally constant* (in short clc) if $H^\bullet S^\bullet$ is locally constant.

(ii) Let $\mathfrak{X} : X_n = X \supset X_{n-1} \supset X_{n-2} \supset \ldots \supset X_1 \supset X_0 \supset X_{-1} = \emptyset$ be a filtration of X by closed subsets. We say that $S^\bullet$ is $\mathfrak{X}$*-cohomologically locally constant* (in short $\mathfrak{X}$-clc) if $H^\bullet S^\bullet$ is locally constant on each stratum $X_i - X_{i-1}$ $(0 \leqq i \leqq n)$. We say that $S^\bullet$ is $\mathfrak{X}$*-cohomologically constructible* (in short $\mathfrak{X}$-cc) if it is $\mathfrak{X}$-clc and the stalks $H^\bullet S^\bullet_x$ $(x \in X)$ are finitely generated.

(iii) $S^\bullet$ is *cohomologically constructible* (in short cc) if it satisfies the following conditions:

CC1. For $x \in X$ and $m \in \mathbb{Z}$, the inverse system $\mathbf{H}^m_c(U_x; S^\bullet)$ (over all open neighborhoods of x) is essentially constant and its limit is finitely generated.

CC2. For $x \in X$ and $m \in \mathbb{Z}$, the direct system $\mathbf{H}^m(U_x; S^\bullet)$ (over all neighborhoods of x) is essentially constant and its limit is finitely generated.

CC3. For $x \in X$ and $m \in \mathbb{Z}$, $H^m(f_x^! S^\bullet) = \varprojlim \mathbf{H}^m_c(U_x; S^\bullet)$, where U_x runs over the open neighborhoods of x and $f_x : x \to X$ is the inclusion.

CC4. (*Property* (P,Q) *of Wilder*). If $P \subset Q$ are open in X, $\bar{P} \subset Q$ and $\bar{P}$

is compact, then the image of $\mathbb{H}_c^j(P;S^\cdot)$ in $\mathbb{H}_c^j(Q;S^\cdot)$ is finitely generated.

If X is a pseudomanifold with a topological stratification $\mathfrak{X}$, we shall show that $S^\cdot$ is cohomologically constructible if it is $\mathfrak{X}$ - cohomologically constructible, and that the complexes of sheaves constructed in paragraph 2 are $\mathfrak{X}$ - cohomologically constructible. Here $\mathfrak{X}$ may be unrestricted (2.1).

3.4 *Remarks.* a) If CC4 holds and X is compact, we can take $P = Q = X$ and we find that $\mathbb{H}^m(X;S^\cdot)$ is finitely generated. This will show in particular that the intersection cohomology groups of compact pseudomanifolds are finitely generated.

b) One can show that the following relations hold :

(i) CC1 $\Longrightarrow$ CC3 ;

(ii) CC1 $\Longrightarrow$ CC4 ;

(iii) if $H^\cdot(S^\cdot_x)$ is finitely generated for all $x \in X$, then CC1 $\Longrightarrow$ CC2 ;

(iv) if $H(f_x^!S^\cdot)$ is finitely generated and R is artinian, then CC2 (at x) $\Longrightarrow$ CC1 (at x).

We shall prove here only (ii) and (iii). We shall use actually only (ii). See 3.17 for some references and further comments.

3.5 THEOREM . CC1 $\Longrightarrow$ CC4 .

The condition CC4 is certainly satisfied for j very large, since in this case $\mathbb{H}_c^j(Q;S^\cdot) = 0$. We can therefore use descending induction on j. We assume that CC4 holds for j + 1 (and any open subsets $P \subset Q$ with $\bar{P}$ compact, $\bar{P} \subset Q$).

Let $E = \{U \subset Q | U \text{ open}, \bar{U} \text{ compact}, \bar{U} \subset Q\}$ and let $E^j = \{U \in E | \mathrm{Im}(\mathbb{H}_c^j(U;S^\cdot) \to \mathbb{H}_c^j(Q;S^\cdot)) \text{ is finitely generated }\}$. We must show that $P \in E^j$. Since R is noetherian, it is clear that if $P \subset V$ for some $V \in E^j$, then $P \in E^j$. It is therefore sufficient to show that every compact subset K of Q has a neighborhood $V \in E^j$.

It follows immediately from CC1 that every $x \in Q$ has a neighborhood which is an element of E^j. If $K \subset Q$ is compact, we can therefore find $U_1 \ldots, U_k \in E^j$ such that $K \subset U_1 \cup \ldots \cup U_k$. We must show that we can do this with k = 1. The crucial case is k = 2. Indeed, if this is settled we can use induction for $k \geqq 3$: the compact set $K - (U_1 \cup \ldots \cup U_{k-2})$ is contained in $U_{k-1} \cup U_k$, hence has a neighborhood $U'_{k-1} \in E^j$, and $K \subset U_1 \cup \ldots \cup U_{k-2} \cup U'_{k-1}$.

So let $k = 2$. We can then find an open neighborhood V_1 of $K - U_2$ such that $\bar{V}_1 \subset U_1$. Then $K \subset V_1 \cup U_2$, and we can find an open neighborhood V_2 of $K - V_1$ such that $\bar{V}_2 \subset U_2$. Then $K \subset V_1 \cup V_2$, and we need only to check that $V_1 \cup V_2 \in E^j$. Consider the diagram

$$\begin{array}{ccccc}
\mathbb{H}^j_c(V_1) \oplus \mathbb{H}^j_c(V_2) & \longrightarrow & \mathbb{H}^j_c(U_1) \oplus \mathbb{H}^j_c(U_2) & \overset{\mu}{\searrow} & \\
\downarrow & & \downarrow \alpha & & \mathbb{H}^j_c(Q) \\
\mathbb{H}^j_c(V_1 \cup V_2) & \overset{\beta}{\longrightarrow} & \mathbb{H}^j_c(U_1 \cup U_2) & \overset{\nu}{\nearrow} & \\
\downarrow & & \downarrow \delta & & \\
\mathbb{H}^{j+1}_c(V_1 \cap V_2) & \overset{\gamma}{\longrightarrow} & \mathbb{H}^{j+1}_c(U_1 \cap U_2) & &
\end{array}$$

where hypercohomology is meant with respect to $S^\bullet$ and the columns are given by the Mayer-Vietoris sequences. We must show that Im $(\nu \circ \beta)$ is finitely generated. By hypothesis Im (μ) = Im $(\nu \circ \alpha)$ is finitely generated. Thus $\nu(\mathrm{Im}\ \beta \cap \mathrm{Im}\ \alpha)$ is finitely generated, and we need only to show that Im $\beta/(\mathrm{Im}\ \beta \cap \mathrm{Im}\ \alpha)$ is finitely generated. But

$$\mathrm{Im}\ (\beta)/(\mathrm{Im}\ \beta \cap \mathrm{Im}\ \alpha) \cong \mathrm{Im}\ (\delta \circ \beta) \subset \mathrm{Im}\ (\gamma) \ ,$$

and Im (γ) is finitely generated by induction hypothesis.

3.6 *Proof of* 3.4 (iii). In this proof, hypercohomology is with respect to $S^\bullet$.

By CC1 we can find a fundamental system $(U_i)_{i \geqq -2}$ of open neighborhoods of x with the following properties: for $i \geqq -1$, $\bar{U}_i \subset U_{i-1}$, $\bar{U}_i$ is compact, and Im $(\mathbb{H}^j_c(U_i) \to \mathbb{H}^j_c(U_{i-1}))$ is finitely generated. We need only to show that the direct system $\mathbb{H}^j(\bar{U}_i;S^\bullet)_{i \geqq 1}$ is essentially constant. As $H^\bullet S^\bullet_x = \varinjlim_i \mathbb{H}^j(\bar{U}_i;S^\bullet)$ is finitely generated, it suffices to show that Im $\mathbb{H}^j(\bar{U}_i;S^\bullet) \to \mathbb{H}^j(\bar{U}_{i+2};S^\bullet))$ is finitely generated $(i \geqq 1)$. It is enough to check this when $i = 1$.

We have a commutative diagram with exact rows:

$$\begin{array}{ccccc}
\mathbb{H}_c^j(U_o) & \longrightarrow & \mathbb{H}^j(\bar{U}_1) & \longrightarrow & \mathbb{H}_c^{j+1}(U_o - \bar{U}_1) \\
\downarrow & & \alpha\downarrow & & \nu\downarrow \\
\mathbb{H}_c^j(U_{-1}) & \xrightarrow{\gamma} & \mathbb{H}^j(\bar{U}_2) & \xrightarrow{\delta} & \mathbb{H}_c^{j+1}(U_{-1} - \bar{U}_2) \\
\lambda\downarrow & & \beta\downarrow & & \downarrow \\
\mathbb{H}_c^j(U_{-2}) & \xrightarrow{\mu} & \mathbb{H}^j(\bar{U}_3) & \longrightarrow & \mathbb{H}_c^{j+1}(U_{-2} - \bar{U}_3)
\end{array}$$

Now Im (λ) is finitely generated by construction and Im (ν) is finitely generated since CC1 implies the (P,Q) - property. It follows that

$$\beta(\mathrm{Im}\ (\alpha) \cap \mathrm{Im}\ (\gamma)) \subset \mathrm{Im}\ (\beta o \gamma) = \mu(\mathrm{Im}\ (\lambda)) \qquad \text{and}$$

$$\mathrm{Im}\ (\alpha)/(\mathrm{Im}\ (\alpha) \cap \mathrm{Im}\ (\gamma)) \cong \mathrm{Im}\ (\delta o \alpha) \subset \mathrm{Im}\ (\nu)$$

are also finitely generated. This implies that Im $(\beta o \alpha)$ is finitely generated, as required.

3.7 PROPOSITION. *Let* M *be a manifold,* m *its dimension,* $\mathcal{S}^\bullet$ *a clc differential graded sheaf on* M *and* $x \in M$. *Then :*
a) *The direct system* $\mathbb{H}^i(U;\mathcal{S}^\bullet)$ *(resp. the inverse system* $\mathbb{H}_c^i(U;\mathcal{S}^\bullet)$*) is constant on the set of neighborhoods of* x *which are homeomorphic to open balls, and equal to* $H^i S_x^\bullet$ *(resp.* $H^{i-m} S_x^\bullet$*)* , $(i \in \mathbb{Z})$.
b) $f_x^! \mathcal{S}^\bullet = f_x^* \mathcal{S}^\bullet[-m]$. *In particular* $H^i(f_x^! \mathcal{S}^\bullet) = H^{i-m}(\mathcal{S}_x^\bullet)$ $(i \in \mathbb{Z})$.
Moreover if $H^\bullet \mathcal{S}^\bullet$ *has finitely generated stalks, then* $\mathcal{S}^\bullet$ *is cc.*

a) $H^\bullet \mathcal{S}^\bullet$ is constant on U and therefore the spectral sequences for $\mathbb{H}(U;\mathcal{S}^\bullet)$ and $\mathbb{H}_c(U;\mathcal{S}^\bullet)$ collapse. We have

$$H^i(U;H^\bullet\mathcal{S}^\bullet) = \begin{cases} H^\bullet\mathcal{S}_x^\bullet & \text{if } i = 0 \\ 0 & \text{if } i \neq 0 \end{cases}$$

(1)

$$H_c^i(U;H^\bullet\mathcal{S}^\bullet) = \begin{cases} H^\bullet\mathcal{S}_x^\bullet & \text{if } i = m \\ 0 & \text{if } i \neq m , \end{cases}$$

and therefore

$$(2)\qquad \mathbf{H}^j(U;S^\bullet) = H^j(S^\bullet_x),\quad \mathbf{H}^j_c(U;S^\bullet) = H^{j-m}(S^\bullet_x) \qquad (j \in \mathbb{Z}) .$$

This implies (a).

b) Let U be an open neighborhood of x whose closure is homeomorphic to a closed ball and let $J^\bullet$ be an injective resolution of $S^\bullet$. We have a natural map

$$(3)\qquad f^!_x S^\bullet = \Gamma_{\{x\}}(U;J^\bullet) \to \Gamma_c(U;J^\bullet) .$$

We claim that the cup product with a fundamental class $[U]_c$ with compact support yields a q.i.

$$(4)\qquad \Gamma(U;J^\bullet) \to \Gamma_c(U;J^\bullet)[m] .$$

This cup product induces a morphism of the hypercohomology spectral sequences which increases the total degree by m . On the other hand, for any R-module E , it gives an isomorphism of $H^i(U;E)$ onto $H^{i+m}_c(U;E)$ for all i (note that both terms are zero for $i \neq 0$ and equal to E for $i = 0$). It follows then from (1) that $\cup\,[U]_c$ yields an isomorphism of the E_2 terms increasing the total degree by m. Together with (2), this implies our assertion. By (a) we know that

$$(5)\qquad \Gamma(U;J^\bullet) \longrightarrow J^\bullet_x \leftarrow S^\bullet_x$$

is a quasi-isomorphism. So we need only to check that (3) yields a quasi-isomorphism. That is, we must show that $H^i(f^!_x S^\bullet) = \mathbf{H}^i_c(U;S^\bullet)$, or equivalently that $H^i(f^!_x S^\bullet) = \mathbf{H}^{i-m}(U;S^\bullet)$ $(i \in \mathbb{Z})$.

We consider the following commutative diagram with exact rows:

$$(6)\qquad \begin{array}{ccccccc}
\cdots \to H^i(f^!_x S^\bullet) & \longrightarrow & \mathbf{H}^i(\bar U;S^\bullet) & \longrightarrow & \mathbf{H}^i(\bar U - \{x\};S^\bullet) \to \cdots \\
\downarrow \alpha_i & & \downarrow \beta_i & & \downarrow \gamma_i \\
\cdots \to \mathbf{H}^i_c(U;S^\bullet) & \longrightarrow & \mathbf{H}^i(\bar U;S^\bullet) & \longrightarrow & \mathbf{H}^i(\bar U - U;S^\bullet) \to \cdots
\end{array}$$

We want to show that α_i is an isomorphism. For this, it suffices to prove that γ_i is an isomorphism for all i's. This map is induced by the restriction map. It suffices to show that it induces an isomorphism of the E_2 - terms of the corresponding hypercohomology spectral sequences i.e. we have to see that the restriction

$$(7) \qquad H^{\bullet}(\bar{U} -\{x\};\mathcal{H}^{\bullet}\mathcal{S}^{\bullet}) \longrightarrow H^{\bullet}(\bar{U} - U;\mathcal{H}^{\bullet}\mathcal{S}^{\bullet}) ,$$

is an isomorphism. But we may write

$$(8) \qquad \bar{U} - \{x\} = S^{n-1} \times (0,1] , \bar{U} - U = S^{n-1} \times \{1\}$$

and the inclusion map is the obvious one, whence our claim.

Suppose now that the stalks of $\mathcal{H}^{\bullet}\mathcal{S}^{\bullet}$ are finitely generated. Part (a) shows then that CC1 and CC2 hold. Part (b) proves CC3 . By (3.5) we have also CC4 . Thus $\mathcal{S}^{\bullet}$ is cohomologically contructible.

3.8 LEMMA. *Let* $(M,\mathfrak{M})$ *be a stratified pseudomanifold,* Y *a topological space and equip* $X = Y \times M$ *with the filtration* $\mathfrak{X}$ *by its closed subsets* $Y \times M_i$. *Let* $\pi : X \to Y$ *be the projection, and for* $y \in Y$ *let* $M_y = \pi^{-1}(y)$. *Let* $\mathcal{S}^{\bullet}$ *be* $\mathfrak{X}$*-clc on* X . *Then*

a) *Suppose that* Y *is a closed ball. Then the natural map* $\mathbb{H}^{\bullet}_c(X;\mathcal{S}^{\bullet}) \to \mathbb{H}^{\bullet}_c(M_y;\mathcal{S}^{\bullet})$ *is an isomorphism for all* $y \in Y$.

b) *Suppose that* M *is compact and that every point in* Y *has a fundamental system of neighborhoods which are homeomorphic to a closed ball. Then* $R\pi_*\mathcal{S}^{\bullet}$ *is clc. If moreover* Y *is contractible, then* $\mathbb{H}^{\bullet}(X;\mathcal{S}^{\bullet}) \to \mathbb{H}^{\bullet}(M_y;\mathcal{S}^{\bullet})$ *is an isomorphism. If* Y *is an open ball of dimension* d , *then* $\mathbb{H}^j_c(X;\mathcal{S}^{\bullet}) = \mathbb{H}^{j-d}(M_y;\mathcal{S}^{\bullet})$.

a) Let $\tau : X \to M_y$ be the projection and $j : M_y \to X$ the inclusion. By definition:

$$\mathbb{H}^{\bullet}_c(M_y;\mathcal{S}^{\bullet}) = \mathbb{H}^{\bullet}_c(M_y;j^*\mathcal{S}^{\bullet}) .$$

We also have

$$\mathbb{H}^{\bullet}_c(X;\mathcal{S}^{\bullet}) = \mathbb{H}^{\bullet}_c(M_y;R\tau_*\mathcal{S}^{\bullet})$$

since τ is proper (1.6). Suppose for simplicity that $\mathcal{S}^{\bullet}$ is injective (if this is not the case, replace $\mathcal{S}^{\bullet}$ by an injective resolution).There is then a natural map $f : R\tau_*\mathcal{S}^{\bullet} = \tau_*\mathcal{S}^{\bullet} \to j^*\mathcal{S}^{\bullet}$ defined as follows.

A section γ of $\tau_* S^\bullet$ over an open subset U of M_y is by definition a section of $S^\bullet$ over $\tau^{-1}(U)$. Then $f(\gamma)$ is the restriction of γ to $\tau^{-1}(U) \cap M_y = U$. It is clear that $j^* : \mathbb{H}^\bullet_c(X;S^\bullet) \to \mathbb{H}^\bullet_c(M_y;S^\bullet)$ corresponds to $f^\bullet : \mathbb{H}^\bullet_c(M_y;R\tau_* S^\bullet) \to \mathbb{H}^\bullet_c(M_y;j^*S^\bullet)$. If suffices therefore only to check that f is a quasi-isomorphism. In fact, we shall prove it in case M is a manifold with the trivial stratification and then establish (a) in general by an inductive argument.

Suppose first that M is a manifold with the trivial stratification. Then $H^\bullet S^\bullet$ is locally constant. Let $U \subset M_y$ be open and homeomorphic to an open ball. Then $\tau^{-1}(U) = Y \times U$ is contractible. As $H^\bullet S^\bullet$ is locally constant, $H^\bullet S^\bullet$ must be constant on $\tau^{-1}(U)$. Thus the spectral sequence for $\mathbb{H}^\bullet(\tau^{-1}(U);S^\bullet)$ collapses and gives $\mathbb{H}^\bullet(\tau^{-1}(U);S^\bullet) = H^\bullet S^\bullet_z$ for any $z \in \tau^{-1}(U)$, in particular for any $z \in U$. But $\mathbb{H}^\bullet(\tau^{-1}(U);S^\bullet) = \mathbb{H}^\bullet(U;R\tau_* S^\bullet)$ Thus, letting U_z run over the open neighborhoods of z in M_y, we get

$$H^\bullet(R\tau_* S^\bullet)_z = \varinjlim \mathbb{H}^\bullet(U_z;R\tau_* S^\bullet) = H^\bullet S^\bullet_z = H^\bullet(j^*S^\bullet)_z .$$

Thus f is a quasi-isomorphism and (a) holds in this case.

We use now induction on k. Let $V_k = M - M_{m-k}$ where $m = \dim M$, and $Z_k = M_k - M_{k-1}$, $U_k = Y \times V_k$, $S_{n-k} = Y \times Z_{m-k}$. (We have identified M_y and M). We have then a commutative diagram with exact rows, where hypercohomology is with respect to $S^\bullet$:

$$\begin{array}{ccccccc}
\longrightarrow \mathbb{H}^i_c(U_k) & \longrightarrow & \mathbb{H}^i_c(U_{k+1}) & \longrightarrow & \mathbb{H}^i_c(S_{n-k}) \longrightarrow \\
\downarrow \alpha_i & & \downarrow \beta_i & & \downarrow \gamma_i \\
\longrightarrow \mathbb{H}^i_c(V_k) & \longrightarrow & \mathbb{H}^i_c(V_{k+1}) & \longrightarrow & \mathbb{H}^i_c(Z_{m-k}) \longrightarrow
\end{array}$$

The γ_i's are isomorphisms by (a) since Z_{m-k} is a manifold. By induction on k we may assume that the α_i's are isomorphisms. The β_i's are then isomorphisms by the 5-lemma. For k = m we get (a).

b) Let $y \in Y$ and B be a neighborhood of y in Y which is a closed ball. Since M is compact (a) gives

$$\mathbb{H}^\bullet(\pi^{-1}(B);S^\bullet) = \mathbb{H}^\bullet(\pi^{-1}(z);S^\bullet) \quad \text{for all } z \in B .$$

It follows immediately that $R\pi_* S^\bullet$ is clc.

Suppose now that Y is contractible. We have $\mathbb{H}^{\cdot}(X;\mathcal{S}^{\cdot}) = \mathbb{H}^{\cdot}(Y;R\pi_*\mathcal{S}^{\cdot})$. As Y is contractible, $R\pi_*\mathcal{S}^{\cdot}$ is cohomologically constant and the spectral sequence for π gives $\mathbb{H}^{\cdot}(X;\mathcal{S}^{\cdot}) = H^{\cdot}(R\pi_*\mathcal{S}^{\cdot})_y$. We need therefore only to check that $H^{\cdot}(R\pi_*\mathcal{S}^{\cdot})_y = \mathbb{H}^{\cdot}(M_y;\mathcal{S}^{\cdot})$. But

$$H^{\cdot}(R\pi_*\mathcal{S}^{\cdot})_y = \varinjlim H^{\cdot}(U_y;R\pi_*\mathcal{S}^{\cdot}) = \varinjlim \mathbb{H}^{\cdot}(\pi^{-1}(U_y);\mathcal{S}^{\cdot}) ,$$

where U_y runs over all neighborhoods of y. For neighborhoods U_y homeomorphic to closed balls, we get by (a) $\mathbb{H}^{\cdot}(\pi^{-1}(U_y);\mathcal{S}^{\cdot}) = \mathbb{H}^{\cdot}(M_y;\mathcal{S}^{\cdot})$, as requested. The second assertion of (b) is proved.

Since M is compact, π is proper and we have also $\mathbb{H}_c^{\cdot}(X;\mathcal{S}^{\cdot}) = \mathbb{H}_c^{\cdot}(Y;R\pi_*\mathcal{S}^{\cdot})$.

If Y is an open ball of dimension d , the spectral sequence for hypercohomology with compact supports gives

$$\mathbb{H}_c^j(Y;R\pi_*\mathcal{S}^{\cdot}) = H^{j-d}(R\pi_*\mathcal{S}^{\cdot})_y$$

for all $y \in Y$. We have checked above that

$$H^{j-d}(R\pi_*\mathcal{S}^{\cdot})_y = \mathbb{H}^{j-d}(M_y;\mathcal{S}^{\cdot}) ,$$

therefore

$$\mathbb{H}_c^j(X;\mathcal{S}^{\cdot}) = \mathbb{H}^{j-d}(M_y;\mathcal{S}^{\cdot}) ,$$ if Y is an open ball if dimension d .

3.9 LEMMA . *Let* $(X,\mathfrak{X})$ *be a stratified pseudomanifold and let* $A^{\cdot} \in \mathrm{DGS}(U_k)$. *Assume that* $A^{\cdot}$ *is* $\mathfrak{X}$ *- clc, and let* $U = B \times \overset{\circ}{c}(L)$ *be a distinguished neighborhood of* $x \in S_{n-k}$. *Then*

a) $\mathbb{H}^{\cdot}(U;Ri_{k*}A^{\cdot}) = \mathbb{H}^{\cdot}(L;A^{\cdot}|_L)$

b) $Ri_{k*}A^{\cdot}$ *is* $\mathfrak{X}$ *- clc on* U_{k+1} .

We have $\mathbb{H}^{\cdot}(U;Ri_{k*}A^{\cdot}) = \mathbb{H}^{\cdot}(U \cap U_k;A^{\cdot})$. But $U \cap U_k = B \times \overset{\circ}{c}(L)^* \cong B \times \mathbb{R} \times L$. By 3.8(b), we get (a), and (b) follows from (a) since U is a distinguished neighborhood of any $y \in B$.

3.10 PROPOSITION . *Let* $(X,\mathfrak{X})$ *be a stratified pseudomanifold,* $\mathcal{S}^\bullet \in DGS(X)$ *be* $\mathfrak{X}$ *- clc and* $x \in X$. *Then:*

a) *The inverse system* $\mathbf{H}_c^j(U;\mathcal{S}^\bullet)$ *is constant over distinguished neighborhoods of* x .

b) *The direct system* $\mathbf{H}^j(U;\mathcal{S}^\bullet)$ *is constant over distinguished neighborhoods* of x.

c) CC3 *holds for* $\mathcal{S}^\bullet$.

d) *For any stratum* Z *of* $\mathfrak{X}$ *, the sheaf* $j_Z^!\mathcal{S}^\bullet$ *is clc on* Z .

e) *If* $\mathcal{S}^\bullet$ *is* $\mathfrak{X}$ *- cc , then* $\mathcal{S}^\bullet$ *is cc.*

a) We use the notation of 2.10 for the stratification. Let k be such that $x \in S_{n-k}$. Let $U = B^{n-k} \times \overset{\circ}{c}(L)$ be a distinguished neighborhood of $x \in S_{n-k}$. Let $Z = U \cap S_{n-k} \cong B^{n-k}$. We have an exact sequence

$$(1) \qquad \longrightarrow \mathbf{H}_c^j(U-Z;\mathcal{S}^\bullet) \to \mathbf{H}_c^j(U;\mathcal{S}^\bullet) \to \mathbf{H}_c^j(Z;\mathcal{S}^\bullet) \longrightarrow$$

Now $\mathcal{S}^\bullet|_{S_{n-k}}$ is clc. Hence by 3.7 , $\mathbf{H}_c^j(Z;\mathcal{S}^\bullet)$ gives a constant inverse system. Since $U - Z = B^{n-k} \times \overset{\circ}{c}(L)^* \cong B^{n-k+1} \times L$, we find by 3.8 (b) that

$$(2) \qquad \mathbf{H}_c^j(U-Z;\mathcal{S}^\bullet) = \mathbf{H}^{j-n+k-1}(L;\mathcal{S}^\bullet) .$$

It follows that $\mathbf{H}_c^j(U-Z;\mathcal{S}^\bullet)$ gives also a constant inverse system. By the 5-lemma, the inverse system $\mathbf{H}_c^j(U;\mathcal{S}^\bullet)$ is constant over distinguished neighborhoods of x . This proves (a) .

c) Let $\bar{U} = \bar{B} \times c(L)$ be a closed distinguished neighborhood of $x \in S_{n-k}$. We have a commutative diagram with exact rows

$$(3) \qquad \begin{array}{ccccccc}
\longrightarrow & H^j(f_x^!\mathcal{S}^\bullet) & \longrightarrow & \mathbf{H}^j(\bar{U};\mathcal{S}^\bullet) & \longrightarrow & \mathbf{H}^j(\bar{U} - \{x\};\mathcal{S}^\bullet) & \longrightarrow \\
 & \downarrow \alpha_j & & \| & & \downarrow \gamma_j & \\
\longrightarrow & \mathbf{H}_c^j(U;\mathcal{S}^\bullet) & \longrightarrow & \mathbf{H}^j(\bar{U};\mathcal{S}^\bullet) & \longrightarrow & \mathbf{H}^j(\bar{U} - U;\mathcal{S}^\bullet) & \longrightarrow
\end{array}$$

We want to prove that α_j is an isomorphism for all j . It is enough to check that γ_j is an isomorphism for all j . Let $d = n - k - 1$. We have

$$(4) \qquad \bar{U} = \bar{B}^{d+1} \times c(L) = c(S^d) \times c(L) = c(S^d * L) ,$$

where S^d is the d-dimensional sphere and $S^d * L$ is the joint of S^d and L . It follows that

$$(5) \qquad \bar{U} - \{x\} = (0,1] \times (S^d * L)$$

$$(6) \qquad \bar{U} - U = \{1\} \times (S^d * L) .$$

We are in the situation of 3.8 (b) with $Y = (0,1]$ and $M = S^d * L$. It follows that γ_j is an isomorphism for all j's.

d) Consider the long exact sequence 1.8(7) in $Sh(S_{n-k})$

$$(7) \qquad \dots \to H^i(j_k^! S^\bullet_{k+1}) \to H^i(j_k^* S^\bullet_{k+1}) \to H^i(j_k^* Ri_{k*} S^\bullet_k) \to \dots$$

The second term is locally constant for each i , and so is the third by 3.9(b). Since S_{n-k} is locally connected, this implies by the 5-lemma that the first term is locally constant for each i .

b) Let $U = B \times \overset{\circ}{c}(L)$ be a distinguished neighborhood of $x \in S_{n-k}$. Consider the long exact sequence (1.8(8)).

$$\to \mathbb{H}^i(B; j_k^! S^\bullet) \to \mathbb{H}^i(U; S^\bullet) \to \mathbb{H}^i(U; Ri_{k*} S^\bullet_k) \to$$

When U runs over a fundamental system of distinguished neighborhoods of x , the first term is constant by (d) and 3.7 , and so it the third one by 3.9(a) . Also the middle one is then constant by the 5-lemma.

e) Assume now $S^\bullet$ to be $\mathfrak{X}$ - cc . To prove (e) we use 3.7 , induction on k and dim X and in particular assume (e) proved for the links. Consequently, CC4 for L implies that $\mathbb{H}^\bullet(L; S^\bullet)$ is finitely generated 3.4(a). Therefore so is $\mathbb{H}^\bullet_c(U - Z; S^\bullet)$ by (2) . But $\mathbb{H}^\bullet(Z; S^\bullet)$ is finitely generated by 3.7 , hence so is $\mathbb{H}^j_c(U; S^\bullet)$ in view of (1). Together with (a), this proves that $S^\bullet$ satisfies CC1 . Then 3.5, 3.6 and (b), (c) show that $S^\bullet$ is cc.

3.11 COROLLARY . (i) *The constant sheaf* $\mathbb{R}_X$ *on* X *is cc.*

(ii) *The assignement* $x \longmapsto H^\bullet(f_x^! S^\bullet)$ *is locally constant on every stratum of* $\mathfrak{X}$.

(iii) *In the situation of 3.9 , the* DGS $Ri_{k*}A^\bullet$ *is* $\mathfrak{X}$ - *cc if* $A^\bullet$ *is* $\mathfrak{X}$ - *cc.*

Proof : (i) follows directly from 3.10 since $\mathbb{R}_X$ is obviously $\mathfrak{X}$ - cc for any $\mathfrak{X}$.

(ii) Let $x \in S_{n-k}$ and denote by g_x the inclusion of x in S_{n-k} . Then $f_x = j_k \circ g_x$ and therefore $f_x^! = g_x^! \circ j_k^!$ (1.9) . By 3.7 , we have then

$$H^i(f_x^! S^\bullet) = H^{i-n+k}(j_k^! S^\bullet)_x \ , \ (i \in \mathbf{Z}) \ ,$$

since $j_k^! S^\bullet$ is clc on S_{n-k} by 3.10(d); (ii) follows.

(iii) We know already that $Ri_{k*}A^\bullet$ is $\mathfrak{X}$ - clc . It remains to check that $H^\bullet(Ri_{k*}A^\bullet)_x$ is finitely generated if $x \in S_{n-k}$. By 3.9(a) this is equal to $\mathbf{H}^\bullet(L;A^\bullet|_L)$. By 3.10(e) $A^\bullet$ is cc and in particular satisfies CC4 . As L is compact $\mathbf{H}^\bullet(L;A^\bullet|_L)$ is therefore finitely generated.

3.12 PROPOSITION . *Let* $(X,\mathfrak{X})$ *be a stratified pseudomanifold and let* $S^\bullet \in$ DGS(X) *satisfy* $(AX1)_{p,\mathfrak{X},E}$ *for some local system* E *on* U_2 . *Then* $S^\bullet$ *is* $\mathfrak{X}$ - *cc and cc* .

In view of 2.5 and 3.10 , it suffices to prove that Deligne's sheaf $P^\bullet(E)$ is $\mathfrak{X}$ - cc . The local system E is $\mathfrak{X}$ - cc on U_2 and $P^\bullet$ is constructed by successive applications of operations of the form Ri_{k*} and $\tau_{\leq p(k)}$, both of which preserve the property to be $\mathfrak{X}$ - cc (by 3.11 (iii) for Ri_{k*} , obviously for $\tau_{\leq p(k)}$) .

3.13 Let M be a topological space and $\pi : X' = X \times M \to X$ the projection. Let $Y \subset X$ and $Y' = \pi^{-1}(Y)$. We have a cartesian diagram

$$(1) \qquad \begin{array}{ccc} Y' & \xrightarrow{\ i'\ } & X' \\ {\scriptstyle \pi'}\downarrow & & \downarrow{\scriptstyle \pi} \\ Y & \xrightarrow{\ i\ } & X \end{array}$$

LEMMA . *Assume* M *is locally contractible* .

(a) *Let* Y *be open and* $S^\cdot \in DGS(Y)$. *Then*

$$(2) \qquad Ri'_*\pi'^*S^\cdot = \pi^*Ri_*S^\cdot \, , \qquad (S^\cdot \in DGS(Y))$$

(b) *Let* Y *be closed and* $T^\cdot \in DGS(X)$.*Then*

$$(3) \qquad \pi'^*i^!T^\cdot = i'^!\pi^*T^\cdot \, .$$

Proof :Let $U \subset X$, $V \subset M$ be open, with V connected, and let $A \in Sh(X)$. Since π^*A is constant on the fibres of π and V is connected, we have

$$(4) \qquad (\pi^*A)\ (U \times V) = A(U) \, .$$

We now prove (a). Using (4) for π and π' , we find that we have for $B \in Sh(Y)$:

$$(5) \qquad i'_*\pi'^*B = \pi^*i_*B \, .$$

In particular, if $S^\cdot \to I^\cdot$ is the canonical flabby resolution of $S^\cdot$ [5] we have

$$(6) \qquad i'_*\pi'^*I^\cdot = \pi^*i_*I^\cdot = \pi^*Ri_*S^\cdot \, .$$

Since π'^* is exact, $\pi'^*S^\bullet \to \pi'^*I^\bullet$ is a resolution. We claim that:

(7) $$\pi'^*I^i \text{ is acyclic for } i^!_* \quad (i \in \mathbb{Z}) .$$

Assume this for the moment. Then $\pi'^*I^\bullet$ can be used to compute $Ri^!_*\pi'^*S^\bullet$, and (6) becomes $Ri^!_*\pi'^*S^\bullet = \pi^*Ri_*S^\bullet$, as claimed.

We need therefore only to prove (7). By construction I^i is a sheaf of the form

(8) $$\prod_{y\in Y} F(y) \text{ , with } F(y) \in Sh(Y) \text{ , } supp(F(y)) \subset \{y\} .$$

It follows from (4) that

(9) $$\pi'^*(\prod_{y\in Y} F(y)) = \prod_{y\in Y} \pi'^*F(y) .$$

It is clear that each $\pi'^*F(y)$ is acyclic for $i^!_*$, since its support is closed in X' . As direct image commutes with direct products, we see that in order to prove (7) it is sufficient to check that if

$\pi'^*F(y) \to J^\bullet(y)$ is an injective resolution $(y \in Y)$, then both $\Pi_{y\in Y}\pi'^* F(y) \to \Pi_{y\in Y}J^\bullet(y)$ and $\Pi_{y\in Y}i^!_*\pi'^*F(y) \to \Pi_{y\in Y}i^!_*J(y)$ are still resolutions. For this, it is enough to prove that if $U \subset Y$, $V \subset M$ are open, with V contractible, then the sequence

(10) $$\prod_{y\in Y} \pi'^*F(y)(U \times V) \to \prod_{y\in Y} J^0(y)(U \times V) \to \prod_{y\in Y} J^1(y)(U \times V) \to \ldots$$

is exact. This can be checked componentwise, and for a fixed $y \in Y$ it amounts to

(11) $$H^i(U \times V;\pi'^*F(y)) = 0 \qquad \text{for } i > 0 .$$

If $y \in U$, then the support of $\pi'^*F(y)|_{U \times V}$ is contained in the contractible space $\{y\} \times V$, and $\pi'^*F(y)|_{\{y\} \times V}$ is a constant sheaf. By 1.11, this implies (11) in this case. But (11) is obvious if $y \notin U$. This completes the proof of (7) and of part (a).

We can prove (b) along the same lines. Using (4), we get

(12) $$\pi'^*\gamma_Y A = \gamma_{Y'}\pi^*A \qquad (A \in Sh(X)) .$$

In order to prove (3) it is therefore sufficient to check that if $T^\bullet \to J^\bullet$ is the canonical resolution of $T^\bullet$ by flabby sheaves [5], then $\pi^* J^i$ is acyclic for $\gamma_{Y'}$ $(i \in \mathbb{Z})$. But each J^i is of the form

$$\text{(13)} \qquad \prod_{x \in X} G(x) \quad \text{with } G(x) \in \mathrm{Sh}(X) \ , \ \mathrm{supp}(G(x)) \subset \{x\} \ .$$

As $\mathrm{supp}(\pi^* G(x))$ is either disjoint from or contained in Y', it is clear that the sheaves $\pi^* G(x)$ are acyclic for $\gamma_{Y'}$. Moreover $\gamma_{Y'}$ commutes with direct products and as in (9) we have $\pi^*(\Pi_{x \in X} G(x)) = \Pi_{x \in X} \pi^* G(x)$. We conclude then as in part (a) that $\pi^* J^i$ is $\gamma_{Y'}$ - acyclic.

Remark.: The same method can be used to prove the following version of the Vietoris-Begle theorem. Assume that M is contractible and locally contractible, and let $T^\bullet \in \mathrm{DGS}(X)$. Then

$$\text{(14)} \qquad \mathbb{H}^\bullet(X \times M; \pi^* T^\bullet) = \mathbb{H}^\bullet(X; T^\bullet) \ .$$

We check first that for any $A \in \mathrm{Sh}(X)$, $\pi^* A$ is acyclic for π_*. By 1.11 (b) this is true for skyscraper sheaves. The same arguments as above show then that it is true for arbitrary products of skyscraper sheaves. If $A \to J^\bullet$ is the canonical flabby resolution, $\pi^* J^\bullet$ can therefore be used to compute $R\pi_* \pi^* A$. But applying π_* to $\pi^* A \to \pi^* J^\bullet$ and using (4), we get $A \to J^\bullet$ back, which is still exact. This proves that $\pi^* A$ is acyclic for π_*. Therefore we have

$$\text{(15)} \qquad R\pi_* \circ \pi^* = \pi_* \circ \pi^* = \mathrm{Id} \ ,$$

and, using 1.6 (3) :

$$\text{(16)} \quad \mathbb{H}^\bullet(X \times M; \pi^* T^\bullet) = \mathbb{H}^\bullet(X; R\pi_* \pi^* T^\bullet) = \mathbb{H}^\bullet(X; \pi_* \pi^* T^\bullet) = \mathbb{H}^\bullet(X; T^\bullet) \ .$$

3.14 LEMMA . *Fix a perversity* p . *Let* $(Y, \mathfrak{Y})$ *be a stratified pseudomanifold,* M *be a manifold and let* $X = M \times Y$ *be equipped with the stratification* $M \times Y \supset M \times Y_{m-2} \supset \ldots \supset M \times Y_{-1} = \emptyset$. *Let* $\pi : X \to Y$ *be the projection and* E *be a local system on* $Y - Y_{m-2}$
$P^\bullet_X(\pi^* E) = \pi^* P^\bullet_Y(E)$.

Let $V_k = Y - Y_k$, let i_k^Y be the inclusion $V_k \to V_{k+1}$ and $\pi_k : U_k \to V_k$ be the projection. We show that $(\mathcal{P}_X^\bullet)_k = \pi_k^*(\mathcal{P}_Y^\bullet)_k$. For $k = 2$ this is clear. Assuming it holds for k , we have :

$$(\mathcal{P}_X^\bullet)_{k+1} = \tau_{\leq p(k)} Ri_{k*} (\mathcal{P}_X^\bullet)_k = \tau_{\leq p(k)} Ri_{k*} \pi_k^* (\mathcal{P}_Y^\bullet)_k =$$

$$= \tau_{\leq p(k)} \pi_{k+1}^* Ri_{k*}^Y (\mathcal{P}_Y^\bullet)_k = \pi_{k+1}^* \tau_{\leq p(k)} Ri_{k*}^Y (\mathcal{P}_Y^\bullet)_k = \pi_{k+1}^* (\mathcal{P}_Y^\bullet)_{k+1}$$

in view of 3.13 and because inverse image commutes with truncation.

3.15 LEMMA. *Let* $(X,\mathfrak{X})$ *be a stratified pseudomanifold and* E *be a local system on* U_2. *Let* $x \in S_{n-k}$ *and* $U = B^{n-k} \times \overset{\circ}{c}(L)$ *be a distinguished neighborhood of* x . *Then*

$$\mathbb{H}^i(U;\mathcal{P}^\bullet(E)) = \begin{cases} \mathbb{H}^i(L;\mathcal{P}_L^\bullet(E)) , & \text{if } i \leq p(k) , \\ 0 & \text{if } i > p(k) , \end{cases}$$

[Here $\mathcal{P}_L^\bullet(E)$ is Deligne's sheaf on L with respect to the local system $E|_{L \cap U_2}$.]

By 3.12 and 3.10 (b), $\mathcal{P}^\bullet$ is $\mathfrak{X}$ - cc and $H^\bullet(\mathcal{P}_x^\bullet) = \mathbb{H}^\bullet(U;\mathcal{P}^\bullet)$. In particular $\mathbb{H}^i(U;\mathcal{P}^\bullet) = 0$ for $i > p(k)$ by condition (b) of (AX1).
As

$$U - B^{n-k} = B^{n-k} \times \overset{\circ}{c}(L)^* \cong B^{n-k+1} \times L ,$$

3.8 (b) and 3.14 show that

$$\mathbb{H}^i(U - B^{n-k};\mathcal{P}^\bullet) = \mathbb{H}^i(L;\mathcal{P}_L^\bullet(E)) .$$

In particular this is constant over distinguished neighborhoods, and we get a commutative diagram

$$\begin{array}{ccc} \mathbb{H}^i(U;\mathcal{P}^\bullet) & \longrightarrow & \mathbb{H}^i(L;\mathcal{P}^\bullet_L) \\ \| & & \| \\ H^i(\mathcal{P}^\bullet_x) & \longrightarrow & H^i(Ri_{k*}\mathcal{P}^\bullet_k)_x \end{array},$$

where the bottom map is given by attachment. By condition (c) of (AX1) it is an isomorphism for $i \leqq p(k)$.

3.16 *Remark on constructibility in* [6] . This basic notion there is what is called here $\mathfrak{X}$ - cc. However some general results involving cc are also needed, which is why that notion is also considered. The theorem on p. 84 of [6] shows that $\mathfrak{X}$ - cc implies CC2, in fact more strongly the constancy of this inductive system. It is further stated there that this implies the other conditions CC1 of 3.3, but I do not see that CC1 follows from CC2 (unless R is artinian). Our 3.8 is related to 1.13 (17) of [6]. There, however, only clc DGS are considered, but in the proof of the lemma in 3.1, p.101, it is used for a DGS which is $\mathfrak{X}$ - cc.

3.17 The above contains all that we need in the sequel about constructibility. For information, we add here a few comments and references.

a) The results mentioned in 3.4 are all proved in [10], Exp. 7.8.

b) Let us say that a direct (resp. inverse) system $\{A_i\}$ is *essentially finitely generated* if, given i, there exists $j \geqq i$ such that $A_i \to A_j$ (resp. $A_j \to A_i$) has a finitely generated image. Consider the four following conditions

(i) For each $x \in X$, the inverse system $\mathbb{H}^\bullet_c(U;\mathcal{S}^\bullet)$ is essentially finitely generated.

(ii) For each $x \in X$, the direct system $\mathbb{H}^\bullet(U;\mathcal{S}^\bullet)$ is essentially finitely generated.

(iii) Property (P,Q).

(iv) If K is compact and contained in the interior of Q, the restriction map $\mathbb{H}^\bullet(Q;\mathcal{S}^\bullet) \to \mathbb{H}^\bullet(K;\mathcal{S}^\bullet)$ has a finitely generated image.

Obviously CC1 (resp. CC2) implies (i) (resp. (ii)). Moreover, the proof of 3.5 actually shows that (i) implies (iii), whence the equivalence of these two conditions. In fact the four conditions are equivalent. This is proved in [4 :p.77-80] when $S^\bullet$ is the constant sheaf R_X but the proof is general. The argument of 3.5 can already be found in [1;2], but in [1] it was directly inspired by an earlier one of R.L.Wilder.

c) Condition (i) for R_X is equivalent to "cohomological local connectedness" in all dimensions, noted clc^∞, defined in [1;2;4]; again, in a different language, it also goes back to R.L.Wilder. Thus if R_X is cc, then X is clc^∞ in the sense of these references.

§ 4 REFORMULATION OF THE AXIOMS AND TOPOLOGICAL INVARIANCE OF IH

In this paragraph X is a pseudomanifold of dimension n . We consider a fixed perversity p , and let q be the dual perversity (i.e. $q(k) = k - p(k) - 2$) .

4.1 Let $\mathfrak{X}$ be a filtration of X by closed subsets

$$X_n = X \supset X_{n-2} \supset X_{n-3} \supset \ldots \supset X_{-1} = \emptyset .$$

As usual, we set :

$$U_k = X - X_{n-k} , \quad S_k = X_k - X_{k-1} ,$$

and let

$$i_k : U_k \to U_{k+1} , \quad j_k : S_{n-k} \to U_{k+1} ,$$

$$f_x : \{x\} \to X \qquad (x \in X)$$

be the inclusions ; unless otherwise stated, we write

$$S_k^\bullet = S^\bullet|_{U_k} \quad \text{if } S^\bullet \in DGS(X) .$$

Let E be a local system on U_2 . In paragraph 2 we considered the following set of conditions on $S^\bullet \in DGS(X)$.

$(AX1)_{\mathfrak{X},E}$:

(1a) (normalization) : *$S^\bullet$ is bounded, $S^i = 0$ for $i < 0$ and $S_2^\bullet = E$.*

(1b) *If $x \in S_{n-k}$, then $H^j(S_x^\bullet) = 0$ for $j > p(k)$.*

(1c) *The attachment map $\alpha_k : S_{k+1}^\bullet \to Ri_{k*}S_k^\bullet$ is a quasi-isomorphism up to $p(k)$.*

We have seen (3.12) that if $\mathfrak{X}$ is a topological stratification, then any $S^\bullet$ satisfying $(AX1)_{\mathfrak{X},E}$ is $\mathfrak{X}$ - cc (hence also $\mathfrak{X}$ - clc).

4.2 We consider now a second set of conditions on $S^\bullet \in DGS(X)$.

$(AX1')_{\mathfrak{X},E}$:

(1'a) *$S^\cdot$ is bounded, $S^i = 0$ for $i < 0$, $S_2^\cdot = E$ and $S^\cdot$ is $\mathfrak{X}$ - clc .*

(1'b) *If $x \in S_{n-k}$, then $H^j(S_x^\cdot) = 0$ for $j > p(k)$.*

(1'c) *If $x \in S_{n-k}$, then $H^j(f_x^! S^\cdot) = 0$ for $j < n - q(k)$.*

Notice that (1b) is the same as (1'b) and that (1'a) is (1a) together with the requirement that $S^\cdot$ be $\mathfrak{X}$ - clc .

4.3 PROPOSITION. *Assume that each stratum S_{n-k} is a manifold of dimension $n - k$ or is empty. Assume also that $S^\cdot \in DGS(X)$ is $\mathfrak{X}$-clc and that $j_k^! S^\cdot$ is clc for $2 \leq k \leq n$. Then $S^\cdot$ satisfies $(AX1)_{\mathfrak{X},E}$ if and only if it satisfies $(AX1')_{\mathfrak{X},E}$.*

We must show that in presence of (1a), (1b) and of the hypotheses of the proposition, the conditions (1c) and (1'c) are equivalent. It is convenient to consider also

(1"c) *If $x \in S_{n-k}$, then $H^j(j_k^! S^\cdot)_x = 0$ for $j \leq p(k) + 1$.*

By 1.8(7), we have for $x \in S_{n-k}$ an exact sequence :

$$\longrightarrow H^j(j_k^! S^\cdot)_x \longrightarrow H^j(S_x^\cdot) \xrightarrow{\alpha_k^j} H^j(Ri_{k*}S_k^\cdot)_x \longrightarrow$$

If $H^j(j_k^! S^\cdot)_x = 0$ for $j \leq p(k) + 1$, the attachment map is certainly an isomorphism for $j \leq p(k)$. Conversely, if α_k^j is an isomorphism for $j \leq p(k)$, then $H^j(j_k^! S^\cdot)_x = 0$ obviously for $j \leq p(k)$, but then also for $j = p(k) + 1$, because $H^{p(k)+1}(S_x^\cdot) = 0$. Thus (1c) $\Longleftrightarrow$ (1"c), modulo (1b).

Let now ℓ_x be the inclusion of x in S_{n-k} . Then $f_x = j_k \circ \ell_x$ and $f_x^! = \ell_x^! \circ j_k^!$. Thus $H^j(f_x^! S^\cdot) = H^j(\ell_x^!(j_k^! S^\cdot))$. By hypothesis, $j_k^! S^\cdot$ is clc . Since S_{n-k} is a manifold of dimension $n - k$, (3.7.b) gives then

$$H^j(\ell_x^!(j_k^! S^\cdot)) = H^{j-n+k}(j_k^! S^\cdot)_x .$$

As $j < n - q(k) \Longleftrightarrow j - n + k \leq p(k) + 1$, we get (1'c) $\Longleftrightarrow$ (1"c). This proves the proposition.

4.4 COROLLARY . *If $\mathfrak{X}$ is a topological stratification, then* $(AX1)_{\mathfrak{X},E} \Longleftrightarrow (AX1')_{\mathfrak{X},E}$.

We know now that $S^\bullet$ is $\mathfrak{X}$ - clc if it satisfies $(AX1)_{\mathfrak{X},E}$ (3.12) or $(AX1')_{\mathfrak{X},E}$. By (3.10.d) , $j_k^! S^\bullet$ is then clc $(2 \leq k \leq n)$.

4.5 COROLLARY . *If $\mathfrak{X}$ is a topological stratification,* $(AX1)_{\mathfrak{X},E}$ *characterizes $S^\bullet$ uniquely up to quasi-isomorphism, and any $S^\bullet$ satisfying* $(AX1')_{\mathfrak{X},E}$ *is $\mathfrak{X}$ - cc* .

This follows from 2.5 and 3.12 .

4.6 Let $S^\bullet \in DGS(X)$, $j \in \mathbb{Z}$. We shall consider the following subsets of X :

(1) $$\{x \in X | H^j(S^\bullet_x) \neq 0\}$$

(2) $$\{x \in X | H^j(f_x^! S^\bullet) \neq 0\}$$

(1) is the support of $H^j(S^\bullet)$. By analogy (2) is sometimes called the j-th homological cosupport of $S^\bullet$. It is also convenient to introduce the following :

Notation: Let p be a perversity. For $j \in \mathbb{N}$ we set $p^{-1}(j) = \min\{c | p(c) \geq j\}$, with $p^{-1}(j) = \infty$ if $j > p(n)$.

It is useful to remember the following rule. For $2 \leq k \leq n$

(3) $$p(k) \geq j \Longleftrightarrow k \geq p^{-1}(j) .$$

4.7 Let $S^\bullet \in DGS(X)$. We consider the following set of conditions on $S^\bullet$.

$(AX2)_{\mathfrak{X},E}$:

$(2a)_{\mathfrak{X}}$ *$S^\bullet$ is bounded , $S^i = 0$ for $i < 0$, $S^\bullet_2 = E$ and $S^\bullet$ is $\mathfrak{X}$ - clc.*

(2b) dim supp $(H^j S^\bullet) \leq n - p^{-1}(j)$ *for all* $j > 0$.

(2c) dim $\{x \in X | H^j(f_x^! S^\bullet) \neq 0\} \leq n - q^{-1}(n - j)$ *for all* $j < n$.

Notice that $(2a)_{\mathfrak{X}}$ is the same as $(1'a)_{\mathfrak{X}}$.

4.8 *Remark*. Here only the first condition depends on $\mathfrak{X}$. This dependence is emphasized because we shall later replace this condition by another one which does not refer to any privileged topological stratification.

4.9 PROPOSITION . *Let* $S^\bullet \in DGS(X)$. *Assume that each stratum* S_{n-k} *is a manifold of dimension* $n-k$ *or is empty, and that* $j_k^! S^\bullet$ *is clc for each* k $(2 \leq k \leq n)$. *Then* $S^\bullet$ *satisfies* $(AX1')_{\mathfrak{X},E}$ *if and only if it satisfies* $(AX2)_{\mathfrak{X},E}$.

Notice first that

$$(1) \qquad j \leq p(k) \iff n-k \leq n - p^{-1}(j) .$$

$$(2) \qquad j \geq n - q(k) \iff n-k \leq n - q^{-1}(n-j).$$

We show that in presence of (1'a) and of the hypothesis of the proposition, (1'b) $\iff$ (2b) and (1'c) $\iff$ (2c) .

(1'b) $\Longrightarrow$ (2b) : If $x \in S_{n-k}$ and $H^j(S_x^\bullet) \neq 0$, then $j \leq p(k)$. Hence $\dim S_{n-k} \leq n-k \leq n - p^{-1}(j)$ by (1) . This implies (2b) .

(2b) $\Longrightarrow$ (1'b) : If $x \in S_{n-k}$ and $H^j(S_x^\bullet) \neq 0$, then $H^j(S_y^\bullet) \neq 0$ for y in some neighborhood of x in S_{n-k} , since $S^\bullet$ is $\mathfrak{X}$ - clc . Therefore $n-k \leq n - p^{-1}(j)$, and $j \leq p(k)$ by (1) . Thus (1'b) holds.

(1'c) $\Longrightarrow$ (2c) : This is proved in the same way as (1'b) $\Longrightarrow$ (2b) .

(2c) $\Longrightarrow$ (1'c) : If $x \in S_{n-k}$ and ℓ_x is the inclusion of x in S_{n-k} , then $f_x = j_k \circ \ell_x$ and $f_x^! = \ell_x^! \circ j_k^!$. In particular $f_x^! S^\bullet = \ell_x^!(j_k^! S^\bullet)$. It follows then from (3.7.b) that $H^j(f_y^! S^\bullet) \cong H^j(f_x^! S^\bullet)$ for y in some neighborhood of x in S_{n-k} . We can then proceed as for (2b) $\longrightarrow$ (1'b).

4.10 COROLLARY. *Assume that* $\mathfrak{X}$ *is a topological stratification. Then* $(AX1')_{\mathfrak{X},E} \iff (AX2)_{\mathfrak{X},E}$. *In particular* $(AX2)_{\mathfrak{X},E}$ *characterizes* $S^\bullet$ *uniquely up to quasi-isomorphism and implies that* $S^\bullet$ *is* $\mathfrak{X}$ - *cc* .

Both axioms require $S^\bullet$ to be $\mathfrak{X}$ - clc . Since $\mathfrak{X}$ is a topological stratification, we know then by 3.10.d that $j_k^! S^\bullet$ is clc for all k $(2 \leq k \leq n)$. The hypotheses of the proposition are fulfilled.

4.11 LEMMA . *Let* M *be a manifold of dimension* n *and let* U *be a dense open subset of* M *whose complement has codimension* ≥ 2 .

a) *If* E, E' *are local systems on* M *and* $f : E|_U \to E'|_U$ *is a morphism then there exists a unique morphism* $g : \mathcal{E} \to \mathcal{E}'$ *which extends* f . *Moreover* g *is an isomorphism if* f *is one.*

b) *If* $\mathcal{E}$ *is a local system on* U *,then there exists a largest open subset* $V \supset U$ *of* M *over which* $\mathcal{E}$ *extends to a local system.*

We can assume that M is connected. Then so is U . [To see this, note that $H^n_c(U;A) = H^n_c(M;A)$ for any coefficients since $H^i_c(M - U;A) = 0$ for $i > n - 2$, and that if A is a field of characteristic two the rank of $H^n_c(U;A)$ is the number of connected components of U .] Let $x \in U$. The category of local systems on M (resp. U) is equivalent to the category of finitely generated $\pi_1(M,x)$-modules (resp. $\pi_1(U,x)$-modules) and the restriction of local systems from M to U corresponds to the restriction of scalars given by $\pi_1(U,x) \to \pi_1(M,x)$.

a) In view of the remarks above, it is sufficient to prove that $\pi_1(U,x) \to \pi_1(M,x)$ is surjective. This follows easily from the fact that for every open ball B in M, the intersection $B \cap U$ is connected, by the above argument.

b) By (a) the local systems $\mathcal{E}'$ which extend $\mathcal{E}$ over open subsets $U' \supset U$ can be glued together.

4.12 Let U, U' be open submanifolds of X whose complements have codimension $\geqq 2$. If $\mathcal{E}$ is a local system on U, we say that $\mathcal{E}$ is defined over U' if $\mathcal{E}|_{U \cap U'}$ can be extended to a local system on U' , or equivalently if U' is contained in the largest open submanifold of X over which $\mathcal{E}$ can be extended to a local system. This extension is also called $\mathcal{E}$.

We say that a stratification of X is adapted to $\mathcal{E}$ if $\mathcal{E}$ is defined over its dense stratum.

4.13 We can now state a new set of conditions on $S^\bullet \in DGS(X)$. Let $\mathcal{E}$ be a local system on some open dense submanifold of X whose complement has codimension $\geqq 2$. Then $(AX2)_{\mathcal{E}}$ consists of the following conditions.

$(AX2)_{\mathcal{E}}$:

(2a) $S^\bullet$ *is bounded,* $S^i = 0$ *for* $i < 0$, $S^\bullet$ *is* $\mathfrak{X}$ - clc *for some topological stratification of* X *and* $S^\bullet|_U = \mathcal{E}|_U$ *for some open dense submanifold* U *of* X *whose complement has codimension* $\geqq 2$ *and over which* $\mathcal{E}$ *is defined.*

(2b) dim supp $H^jS^{\bullet} \leq n - p^{-1}(j)$ *for all* $j > 0$.

(2c) dim $\{x \in X | H^j(f_x^! S^{\bullet}) \neq 0\} \leq n - q^{-1}(n - j)$ *for all* $j < n$.

Notice that the second and the third conditions are the same as for $(AX2)_{\mathfrak{X},E}$.

4.14 *Remarks.* a) Unlike the previous axioms, this one does not refer to any particular stratification of X.

b) Let $S^{\bullet}$ satisfy $(AX2)_E$. In (2a) the open subset U and the stratification $\mathfrak{X}$ are not assumed to be related. We can however certainly assume that $U \subset U_2$. Now $H^0(S^{\bullet})|_U = E$ and $H^i(S^{\bullet})|_U = 0$ for $i \neq 0$. On the other hand $H^i(S^{\bullet})$ is locally constant on U_2 . It follows that $H^0(S^i)|_{U_2} = E$ and $H^i(S^{\bullet})_{U_2} = 0$ for $i \neq 0$. Therefore E is defined over U_2 and $S^{\bullet}|_{U_2} = E$. In particular $S^{\bullet}$ satisfies $(AX2)_{\mathfrak{X},E}$.

As a consequence $S^{\bullet}$ satisfies $(AX2)_E$ if and only if it satisfies $(AX2)_{\mathfrak{X},E}$ for some topological stratification $\mathfrak{X}$ adapted to E .

c) If a local system E is given on a dense open submanifold U of X whose complement has codimension ≥ 2 , there does not necessarily exist $S^{\bullet} \in DGS(X)$ satisfying $(AX2)_E$. For example let $X = \mathbb{R}^2$ and U be the complement of $\{(\frac{1}{n},0) | n \in \mathbb{N}, n \neq 0\} \cup \{(0,0)\}$. It is easy to construct a local system E on U which cannot be extended to a local system on a larger open subset of X . As U does not contain the open stratum of any topological stratification, no $S^{\bullet} \in DGS(X)$ can satisfy $(AX2)_E$.

4.15 THEOREM . *Let E be a local system on some open dense submanifold of* X *whose complement has codimension* ≥ 2 *and let* $\tilde{U}_2$ *be the largest open submanifold of* X *over which* E *extends to a local system. Assume that there exists a topological stratification of* X *which is adapted to* E . *Then there exists* $\tilde{P}^{\bullet} \in DGS(X)$ *satisfying* $(AX2)_E$ *with* $\tilde{P}^{\bullet}|_{\tilde{U}} = E$ *and* $(AX2)_{\mathfrak{X},E}$ *for every topological stratification* $\mathfrak{X}$ *adapted to* E .

We construct $\tilde{P}^{\bullet}$ as the Deligne sheaf associated to the local system E on $\tilde{U}_2$ and a filtration $\tilde{\mathfrak{X}}$ of X by closed subsets

(1) $$\tilde{X}_n = X \supset \tilde{X}_{n-2} \supset \ldots \supset \tilde{X}_{n-k} \supset \ldots \supset \tilde{X}_0 \supset \tilde{X}_{-1} ,$$

with $\tilde{X}_{n-2} = X - \tilde{U}_2$. Here the main difficulty is to construct $\tilde{\mathfrak{X}}$. This will be done by induction on k . We require this filtration to have suitable properties. Let $\tilde{U}_k = X - \tilde{X}_{n-k}$, $\tilde{S}_{n-k} = \tilde{X}_{n-k} - \tilde{X}_{n-k-1}$, with the inclusions $\tilde{i}_k : \tilde{U}_k \to \tilde{U}_{k+1}$, $\tilde{j}_k : \tilde{S}_{n-k} \to \tilde{U}_{k+1}$. We have for each k a differential graded sheaf $\tilde{P}_k$ on $\tilde{U}_k$ defined by $\tilde{P}^{\bullet}_2 = E$ and $\tilde{P}^{\bullet}_{k+1} = \tau_{\leq p(k)} R\tilde{i}_{k*}\tilde{P}^{\bullet}_k$. We want to have for each k $(2 \leq k \leq n)$:

(i) $\tilde{S}_{n-k}$ *is a manifold of dimension* $n-k$ *or is empty* .

(I_k) (ii) $\tilde{j}^{*}_k\tilde{P}^{\bullet}_{k+1}$ *is clc* .

(iii) $\tilde{j}^{!}_k\tilde{P}^{\bullet}_{k+1}$ *is clc* .

(II_k) *For every topological stratification* $\mathfrak{X}$ *of* X *which is adapted to* E , $\tilde{S}_{n-k}$ *is a union of connected components of strata of* $\mathfrak{X}$ *and* $U_{k+1} \subset \tilde{U}_{k+1}$.

Suppose that $\tilde{\mathfrak{X}}$ has these properties. By hypothesis there exists at least one topological stratification $\mathfrak{X}$ of X which is adapted to E . By (II_n) we have then $\tilde{U}_{n+1} \supset U_{n+1} = X$. Let $\tilde{P}^{\bullet} = \tilde{P}^{\bullet}_{n+1} \in DGS(X)$. By construction $\tilde{P}^{\bullet}$ satisfies $(AX1)_{\tilde{\mathfrak{X}},E}$. By (4.3), (4.9) and (I) it satisfies also $(AX2)_{\tilde{\mathfrak{X}},E}$. If $\mathfrak{X}$ is a topological stratification of X which is adapted to E , we find then by (II) that $\tilde{P}^{\bullet}$ is $\mathfrak{X}$-clc . Therefore it satisfies $(AX2)_{\mathfrak{X},E}$. As there exists at least one such stratification, $\tilde{P}^{\bullet}$ satisfies $(AX2)_E$. Thus $\tilde{P}^{\bullet}$ has the required properties.

It is therefore sufficient to construct a stratification $\tilde{\mathfrak{X}}$ satisfying (I_k) and (II_k) for all $k \geq 2$.

The dense stratum $\tilde{U}_2$ is already defined, and $\tilde{P}^{\bullet}_2 = E$. We check first that $\tilde{U}_2$ is a union of connected components of strata of $\mathfrak{X}$, if $\mathfrak{X}$ is a topological stratification of X which is adapted to E . Let $x \in X$ and let $U = B \times \overset{\circ}{c}(L)$ be a distinguished neighborhood of x with respect to $\mathfrak{X}$. It is clear that if x has a neighborhood which is homeomorphic to an open ball, then every $y \in B$ has such a neighborhood. Assume now that U is a manifold. Let $\pi : U \to \overset{\circ}{c}(L)$ be the projection. By induction on the codimension of the stratum containing x , we can assume that E is defined over $B \times \overset{\circ}{c}(L)$. Now if E is defined in some neighborhood of x, then $\pi^*(E|_{\{x\} \times \overset{\circ}{c}(L)})$ is a local system on U which restricts to E on $U \cap U_2$. Thus $B \subset \tilde{U}_2$ if $x \in \tilde{U}_2$. It follows easily that $\tilde{U}_2$ is a union of components of strata of $\mathfrak{X}$.

Suppose now by induction that $\tilde{U}_2,\ldots,\tilde{U}_k$ are already defined and that (I_i), (II_i) hold for $2 \leqq i < k$ (notice that $\tilde{\mathcal{P}}^\bullet_2,\ldots,\tilde{\mathcal{P}}^\bullet_k$ are also defined). Let $\bar{i}_k : \tilde{U}_k \to X$, $\bar{j}_k : \tilde{X}_{n-k} \to X$ be the inclusions and let $\bar{\mathcal{P}}^\bullet_{k+1} = \tau_{\leqq p(k)} R\bar{i}_{k*}\tilde{\mathcal{P}}^\bullet_k$. Let $\tilde{S}'_{n-k}$ be the largest submanifold of $\tilde{X}_{n-k}$ of dimension $n-k$. Let $\tilde{S}''_{n-k}$ (resp. $\tilde{S}'''_{n-k}$) be the largest open subset of $\tilde{X}_{n-k}$ over which $\bar{j}^*_k\bar{\mathcal{P}}^\bullet_{k+1}$ (resp. $\bar{j}^!_k\bar{\mathcal{P}}^\bullet_{k+1}$) is clc.
We take

$$\tilde{S}_{n-k} = \tilde{S}'_{n-k} \cap \tilde{S}''_{n-k} \cap S'''_{n-k} \ , \ \tilde{U}_{k+1} = \tilde{U}_k \cup \tilde{S}_{n-k}$$

It is clear that $\tilde{U}_{k+1}$ is open and satisfies (I_k) . It remains to check that (II_k) holds too. This is asserted by :

4.16 LEMMA. *In this situation, let* $\mathfrak{X}$ *be a topological stratification of* X *adapted to* E *and let* Y *be a connected component of some stratum of* $\mathfrak{X}$. *Then each of the sets* $\tilde{S}'_{n-k}$, $\tilde{S}''_{n-k}$, $\tilde{S}'''_{n-k}$ *has an intersection with* Y *which is either empty or equal to* Y . *Moreover* $Y \subset \tilde{S}'_{n-k} \cap \tilde{S}''_{n-k} \cap \tilde{S}'''_{n-k}$ *if* $\operatorname{codim}_X Y = k$ *and* $Y \not\subset \tilde{U}_k$.

We may assume that $Y \subset \tilde{X}_{n-k}$. The statements concerning $\tilde{S}'_{n-k}$ follow immediately from the existence of distinguished neighborhoods. Let now $x \in Y$ and let $U' = B \times \overset{\circ}{c}(L)$ be a distinguished neighborhood of x in X. Let $\pi : U' \to \overset{\circ}{c}(L)$ be the projection and let $\tilde{U}'_j = U' \cap \tilde{U}_j$, $V_j = \pi(\tilde{U}'_j)$ $(2 \leqq j \leqq k)$. Let also $\tilde{U}'_{k+1} = U'$, $V_{k+1} = \overset{\circ}{c}(L)$.
Then $\tilde{U}'_j = \pi^{-1}(V_j)$. For $2 \leqq j \leqq k$ we have a cartesian square :

$$\begin{array}{ccc} \tilde{U}'_j & \xrightarrow{h_j} & \tilde{U}'_{j+1} \\ {\scriptstyle \pi_j}\downarrow & & \downarrow{\scriptstyle \pi_{j+1}} \\ V_j & \xrightarrow{\bar{h}_j} & V_{j+1} \end{array}$$

where h_j , $\bar{h}_j$ are inclusions and π_j , π_{j+1} are obtained by restricting π . Let $\mathcal{Q}^\bullet_2 = E|_{V_2}$, where V_2 is identified with $\{x\} \times V_2 \subset \tilde{U}_2$. Then $\tilde{\mathcal{P}}^\bullet_2|_{\tilde{U}'_2} = \pi^*_2\mathcal{Q}^\bullet_2$. Define now by induction $\mathcal{Q}^\bullet_3,\ldots,\mathcal{Q}^\bullet_{k+1}$ by

$\mathcal{Q}^\bullet_{j+1} = \tau_{\leqq p(j)} R\bar{h}_{j*}\mathcal{Q}^\bullet_j$. By 3.13 , we have then over U'_3

$$\tilde{P}^\bullet_3 = \tau_{\leqq p(2)} Rh_{2*}\tilde{P}^\bullet_2 = \tau_{\leqq p(2)} Rh_{2*}\pi^*_2\mathcal{Q}^\bullet_2 =$$

$$= \tau_{\leqq p(2)}\pi^*_3 R\bar{h}_{2*}\mathcal{Q}^\bullet_2 = \pi^*_3\tau_{\leqq p(2)} R\bar{h}_{2*}\mathcal{Q}^\bullet_2 = \pi^*_3\mathcal{Q}^\bullet_3 ,$$

and in a similar way we find by induction that

$$\tilde{P}^\bullet_j|_{\tilde{U}'_j} = \pi^*_j\mathcal{Q}^\bullet_j \quad \text{for } 2 \leqq j \leqq k \quad \text{and } \bar{P}^\bullet_{k+1}|_{U'} = \pi^*\mathcal{Q}^\bullet_{k+1} .$$

Consider the inclusions $g : (U' - \tilde{U}'_k) \to U'$ and $\bar{g} : (\overset{\circ}{c}(L) - V_k) \to \overset{\circ}{c}(L)$.

We have a cartesian square

$$\begin{array}{ccc} U' - \tilde{U}'_k & \xrightarrow{\ g\ } & U' \\ \downarrow{\scriptstyle \pi'} & & \downarrow{\scriptstyle \pi} \\ \overset{\circ}{c}(L) - V_k & \xrightarrow{\ \bar{g}\ } & \overset{\circ}{c}(L) \end{array}$$

where π' is the restriction of π. We have then

(1) $$g^*\bar{P}^\bullet_{k+1} = g^*\pi^*\mathcal{Q}^\bullet_{k+1} = \pi'^*(\bar{g}^*\mathcal{Q}^\bullet_{k+1})$$

and by 3.13

(2) $$g^!\bar{P}^\bullet_{k+1} = g^!\pi^*\mathcal{Q}^\bullet_{k+1} = \pi'^*(\bar{g}^!\mathcal{Q}^\bullet_{k+1})$$

From (1) (resp. (2)) it follows that the neighborhood B of x in Y is either entirely contained in $\tilde{S}''_{n-k}$ (resp. $\tilde{S}'''_{n-k}$) or disjoint from it. Therefore $Y \cap \tilde{S}''_{n-k}$ (resp. $Y \cap \tilde{S}'''_{n-k}$) is both open and closed in Y . Since Y is connected, this yields the first assertion pertaining to $\tilde{S}''_{n-k}$ (resp. $\tilde{S}'''_{n-k}$).

If $\operatorname{codim}_X Y = k$, then $\overset{\circ}{c}(L) - V_k$ is a point and $\bar{g}^*\mathcal{Q}^\bullet_{k+1}$, $\bar{g}^!\mathcal{Q}^\bullet_{k+1}$ are automatically clc . By (1) and (2) we find then that $Y \subset \tilde{S}''_{n-k} \cap \tilde{S}'''_{n-k}$. This proves the lemma.

4.17 COROLLARY . *Let* $S^{\cdot}$ *satisfy* $(AX2)_E$. *Then* $S^{\cdot}$ *is quasi-isomorphic to* $\tilde{P}^{\cdot}$.

By 4.14.b, we know that $S^{\cdot}$ satisfies $(AX2)_{\mathfrak{X},E}$ for some topological stratification $\mathfrak{X}$ of X which is adapted to E . The differential graded sheaf $\tilde{P}^{\cdot}$ constructed in 4.15 satisfies also $(AX2)_{\mathfrak{X},E}$. By 4.10, $S^{\cdot} = \tilde{P}^{\cdot}$.

4.18 COROLLARY . *Let* $\mathfrak{X}$ *be a topological stratification of* X , *let* $\mathcal{O}$ *be the orientation sheaf on* U_2 *and* $P^{\cdot} = P^{\cdot}_{\mathfrak{X}}(\mathcal{O})$. *Then* $P^{\cdot}$ *is independent on the choice of* $\mathfrak{X}$.

This is a special case of 4.17 . Notice that every topological stratification of X is adapted to $\mathcal{O}$.

4.19 COROLLARY . *Let* X *be a PL-pseudomanifold. Then the intersection homology groups* $I_pH_i(X;R)$ *defined in* (I) *are independent of the PL-stratification used in their definition.*

This follows from 2.9 and 4.18 .

4.20 *Remarks* on [6] . In our discussion of the axioms, there are some minor differences with [6] on which I would like to make some comments.

First of all, the various sets of conditions [AX i] in [6] always include constructibility assumptions, more precisely what has been called here $\mathfrak{X}$ - cc . They were absent from our version of AX1 in § 2 , but we saw that they follow from it. In AX1' and the various AX2 , we have required $\mathfrak{X}$ - clc , i.e. we have not assumed finite generation of the stalk cohomology . (Of course, we get it back, once the equivalence with AX1 is established, as was pointed out). There are two reasons for this. One is that it may be of interest for the applications to weaken the constructibility assumptions as much as possible. In fact, according to M. Goresky and R. MacPherson, it is an open question as to whether one could also dispense entirely with them in AX2. But there is also a technical reason of more immediate interest in connection with the proof of topological invariance : in the construction of $\tilde{X}$ and $\tilde{P}^{\cdot}$ in 4.15 , and in the similar step in 4.2 of [6] , what comes out naturally from the proof is $\mathfrak{X}$ - clc but not the finite generation of the stalk cohomology. In fact, as far as I can see, the latter is not proved at all in [6].

In the inductive construction of the p-filtration in [6], use is made of cohomology manifolds and the strata are only assumed to be cohomology manifolds. This yields a stratification which may be coarser than $\tilde{\mathfrak{X}}$. We have preferred to stick to manifolds because the proof uses 3.7 , which has been proved here only for manifolds, and also incidentally to show that the recourse to cohomology manifolds is not needed to establish the topological invariance.

In the construction of the analogue $\tilde{P}_p$ of $\tilde{P}$ in [6] it is proved that $\tilde{P}_p$ is clc on the strata of $\tilde{\mathfrak{X}}$. No other reason is given to make sure that the new strata are unions of connected components of old strata. It seems to me that some argument, as given in 4.15 or 4.16, is needed to settle that point.

§ 5 THE DERIVED CATEGORY OF SHEAVES

So far, we have not formally introduced the derived category of Sh(X), although we have made much use of the notion of quasi-isomorphism, often written as an equality, which should really be understood as an isomorphism "in the derived category", but there was hardly any need to be more explicit about it. However, the formalism of derived categories will enter in a more substantial way in the discussion of Verdier biduality and of pairings in § 9 , whence the need of a more formal presentation. We shall limit it however to the needs of this seminar, and refer to [7;12] for a much more thorough and more general treatment.

From Sh(X) we form the category K(ShX) or simply K(X) whose objects are the complexes of sheaves of R-modules on the space X and whose maps are the homotopy classes of morphisms. If X is a point it reduces to the category of complexes of R-modules. It has a translation functor, namely the automorphism which assigns to $(S^\bullet,d)$ the complex $(S^\bullet[1],-d)$. Much of the following would be valid for the category of complexes on an abelian category with enough injectives. (See [7;12].)

A. *Triangles and long exact sequences.*

We are all accustomed to derive long exact sequences in cohomology from short exact sequences of complexes. However we cannot in general replace an element in a short exact sequence in K(X) by an isomorphic one in K(X) (i.e. a homotopy equivalent one) and still get a short exact sequence. So this notion does not make good sense in K(X) , which is not an abelian category; it is even worse if we want to replace an element by a q.i. one, i.e. to work in the derived category. It turns out that the adequate substitutes for short exact sequences to generate long exact cohomology sequences are the so-called *distinguished* (or exact) *triangles*. We shall therefore first introduce those and discuss some of their properties.

5.1 A *triangle* $(A^\bullet,B^\bullet,C^\bullet,u,v,w)$ in K(X) is a sextuple

(1) $$A^\bullet \xrightarrow{u} B^\bullet \xrightarrow{v} C^\bullet \xrightarrow{w} A^\bullet[1]$$

of elements $\mathcal{A}^{\bullet}$, $\mathcal{B}^{\bullet}$, $\mathcal{C}^{\bullet}$ and morphisms u,v,w. It is also written

$$(2)\qquad \begin{array}{ccc} \mathcal{A}^{\bullet} & \xrightarrow{u} & \mathcal{B}^{\bullet} \\ {\scriptstyle w}\nwarrow & & \swarrow{\scriptstyle v} \\ & \mathcal{C}^{\bullet} & \end{array}$$

with usually [1] affixed to $\nwarrow$, to indicate that w is a morphism from $\mathcal{C}^{\bullet}$ to $\mathcal{A}^{\bullet}[1]$.

A morphism of triangles

$$(3)\qquad \begin{array}{ccc} \mathcal{A}^{\bullet} & \xrightarrow{u} & \mathcal{B}^{\bullet} \\ {\scriptstyle w}\nwarrow & & \swarrow{\scriptstyle v} \\ & \mathcal{C}^{\bullet} & \end{array} \longrightarrow \begin{array}{ccc} \mathcal{A}'^{\bullet} & \xrightarrow{u'} & \mathcal{B}'^{\bullet} \\ {\scriptstyle w'}\nwarrow & & \swarrow{\scriptstyle v'} \\ & \mathcal{C}'^{\bullet} & \end{array}$$

is given by a (homotopy) commutative diagram

$$(4)\qquad \begin{array}{ccccccc} \mathcal{A}^{\bullet} & \xrightarrow{u} & \mathcal{B}^{\bullet} & \xrightarrow{v} & \mathcal{C}^{\bullet} & \xrightarrow{w} & \mathcal{A}^{\bullet}[1] \\ \downarrow{\scriptstyle f} & & \downarrow{\scriptstyle g} & & \downarrow{\scriptstyle k} & & \downarrow{\scriptstyle f[1]} \\ \mathcal{A}'^{\bullet} & \xrightarrow{u'} & \mathcal{B}'^{\bullet} & \xrightarrow{v'} & \mathcal{C}'^{\bullet} & \xrightarrow{w'} & \mathcal{A}'^{\bullet}[1] \end{array}$$

It is an isomorphism if f,g,k are homotopy equivalences.

5.2 The *mapping cone* $\mathcal{C}^{\bullet}_u$ of a morphism $u : \mathcal{A}^{\bullet} \to \mathcal{B}^{\bullet}$ in DGS(X) is defined as

$$(1)\qquad \mathcal{C}^{\bullet}_u = \mathcal{A}^{\bullet}[1] \oplus \mathcal{B}^{\bullet} ,$$

with the differential

$$(2)\qquad d(a,b) = (-da, u(a)+db), \quad (a \in \mathcal{A}^{i+1};\ b \in \mathcal{B}^{i};\ i \in \mathbf{Z})$$

The maps :

$$(3)\qquad \begin{aligned} \tilde{v} &: \mathcal{B}^{\bullet} \to \mathcal{C}^{\bullet}_u \text{ , given by } \longmapsto (0,b) \\ \tilde{w} &: \mathcal{C}^{\bullet}_u \to \mathcal{A}^{\bullet}[1] \text{ , given by } (a,b) \longmapsto a \end{aligned}$$

are easily seen to be morphisms in DGS(X), whence a triangle

$$A^\bullet \xrightarrow{u} B^\bullet \xrightarrow{\tilde{v}} C^\bullet_u \xrightarrow{\tilde{w}} A^\bullet[1] \tag{4}$$

to be called a *standard triangle*. If u, u' $\in \mathrm{Hom}_{DGS(X)}(A^\bullet,B^\bullet)$ are homotopic morphisms, then $C^\bullet_u$ and $C^\bullet_{u'}$ are easily seen to be isomorphic in K(X). However the isomorphism is not unique in general. With that proviso $C^\bullet_u$ is a well defined object in K(X). A *distinguished triangle* in K(X) is one which is isomorphic to a standard one.

If $(A^\bullet,B^\bullet,C^\bullet, u,v,w)$ is a triangle, the maps u,v,w induce a long sequence of homomorphisms of derived sheaves

$$\dots \longrightarrow H^i(A^\bullet) \longrightarrow H^i(B^\bullet) \longrightarrow H^i(C^\bullet) \longrightarrow H^{i+1}(A^\bullet) \rightarrow \dots \tag{5}$$

as well as an analogous sequence for hypercohomology. If $(A^\bullet,B^\bullet,C^\bullet,u,v,w)$ is distinguished, these sequences are exact. To see this, it suffices to consider a standard triangle. But then

$$0 \longrightarrow B^\bullet \xrightarrow{v} C^\bullet_u \xrightarrow{w} A^\bullet[1] \longrightarrow 0 \tag{6}$$

is exact. It is easily checked that the long exact sequence of derived sheaves associated to (6) coincides with (5)(in particular the connecting homomorphism for (6) coincides with the map induced by u). For hypercohomology, let $\tilde{u} : J^\bullet_{A^\bullet} \to J^\bullet_{B^\bullet}$ be a map of injective resolutions which extends u . Then $C^\bullet_{\tilde{u}}$ is an injective resolution of $C^\bullet_u$, and the short exact sequence

$$0 \longrightarrow J^\bullet_{B^\bullet} \longrightarrow C^\bullet_{\tilde{u}} \longrightarrow J^\bullet_{A^\bullet}[1] \longrightarrow 0 \tag{7}$$

gives a long exact sequence for hypercohomology, which again coincides with the sequence induced by u,v,w .

5.3 For $A^\bullet \in K(X)$, we shall denote by $C^\bullet_{A^\bullet}$ the cone over the identity map of $A^\bullet$. It is already clear from the long exact sequence that $H^\bullet C^\bullet_{A^\bullet} = 0$, but in fact, $C^\bullet_{A^\bullet}$ is isomorphic to 0 in K(X), i.e. the identity map of $C^\bullet_{A^\bullet}$ is homotopic to zero; it is indeed readily checked that the map h defined by

$$h(a,a') = (a',0) \qquad (a \in A^{i+1} ;\ a' \in A^i) ,$$

provides such a homotopy.

5.4 *The standard triangle associated to a short exact sequence.* Let

(1) $$0 \longrightarrow A^{\bullet} \xrightarrow{u} B^{\bullet} \xrightarrow{v} C^{\bullet} \longrightarrow 0 ,$$

be a short exact sequence of DGS. As already mentioned, distinguished triangles are substitutes for short exact sequences. We want to replace (1) by the standard triangle

(2) 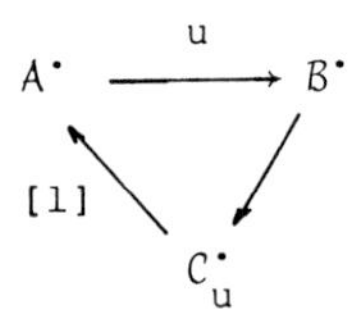

Both (1) and (2) induce long exact sequences. To compare them, we define $m : C^{\bullet}_{u} \to C^{\bullet}$ by

(3) $$m(a,b) = v(b) .$$

LEMMA (i) *The map* m *is a q.i. and up to sign* $Id_{A^{\bullet}}$, $Id_{B^{\bullet}}$ *and* m *induce an isomorphism between the long exact sequences of derived sheaves associated to* (1) *and* (2) .

(ii) *If* (1) *is split, then* m *is an isomorphism in* K(X) .

It is easily seen that m is a morphism and that up to sign $Id_{A^{\bullet}}$, $Id_{B^{\bullet}}$ and m induce a morphism of long exact sequences. Using the 5-lemma, we get (i).

Assume now (1) to be split and let $s : C^{\bullet} \to B^{\bullet}$ be a splitting, (i.e. a R-linear map preserving degrees but not necessarily commuting with the differentials). We have v o s = Id. Since v commutes with d, it is obvious that for $c \in C^{i}$ we have v(sdc - dsc) = 0, i.e. $dsc - sdc \in A^{i+1}$. Therefore

$$n : c \longmapsto (sdc - dsc,\ sc)$$

defines a linear map preserving degrees of $C^{\bullet}$ into $C^{\bullet}_{u}$. It is readily checked to be a morphism. Clearly m o n = Id . We have a direct sum decomposition

$$\mathcal{B}^\bullet = u(\mathcal{A}^\bullet) \oplus s(\mathcal{C}^\bullet)$$

from which we see that

$$\ker m = \mathcal{C}^\bullet_{\mathcal{A}^\bullet} = \mathcal{A}^\bullet[1] \oplus \mathcal{A}^\bullet \subset \mathcal{A}^\bullet[1] \oplus \mathcal{B}^\bullet = \mathcal{C}^\bullet_u$$

whence a direct sum decomposition of complexes

$$\mathcal{C}^\bullet_u = n(\mathcal{C}^\bullet) \oplus \mathcal{C}^\bullet_{\mathcal{A}^\bullet}.$$

But $\mathcal{C}^\bullet_{\mathcal{A}^\bullet} \approx 0$ by 5.3 . Therefore the projection of $\mathcal{C}^\bullet_u$ onto $n(\mathcal{C}^\bullet)$ is a homotopy equivalence and it follows that $n \circ m$ is homotopic to Id. This proves (ii).

Remark . An advantage of the triangle (2) on the exact sequence (1) is that it remains a standard triangle if we apply to it an additive functor $F : Sh(X) \to A$, where A is an abelian category. Moreover the long exact sequences deduced from (2) are read directly from the triangle; that is, the somewhat mysterious connecting homomorphism is now induced by a honest morphism of DGS. Of course we would like to replace $\mathcal{C}^\bullet_u$ by $\mathcal{C}^\bullet$ in (2). Part (ii) of the lemma says that this is possible if (1) is split, and the introduction of the derived category will allow one to do it in general, at the cost however of a more delicate notion of morphisms.

5.5 To illustrate the use of cones, we describe here briefly some variations on the previous way to get long exact sequences, which show so to say that a cone may be viewed as a substitute not only for a quotient, as in 5.4, but also for a kernel. This will not be needed in the sequel.

Since the cone $\mathcal{C}^\bullet_{\mathcal{A}^\bullet}$ on the identity map of $\mathcal{A}^\bullet \in K(X)$ is isomorphic to zero (5.3) we see that the projection of $\mathcal{B}'^\bullet = \mathcal{B}^\bullet \oplus \mathcal{C}^\bullet_{\mathcal{A}^\bullet}$ onto $\mathcal{B}^\bullet$, with kernel $\mathcal{C}^\bullet_{\mathcal{A}^\bullet}$, is an isomorphism. The map $u' : \mathcal{A}^\bullet \to \mathcal{B}'^\bullet$ given by $u'(a) = (u(a),a)$ is then a direct morphism and it is easily seen that $\mathcal{B}'^\bullet/u(\mathcal{A}^\bullet) \cong \mathcal{C}^\bullet_u$, whence a split short exact sequence

$$(1) \qquad 0 \to \mathcal{A}^\bullet \xrightarrow{u'} \mathcal{B}'^\bullet \underset{s}{\overset{v'}{\rightleftarrows}} \mathcal{C}^\bullet_u \to 0$$

where

$u'(a) = (u(a),0,a), v'(b,\tilde{a},a) = (-\tilde{a},b-u(a)),\ s(a,b) = (b,-a,0)$
for $a \in A^i$, $b \in \mathcal{B}^i$ and $\tilde{a} \in A^{i+1}$, which (up to sign) yields the same long exact sequence as 5.2(5).

If u is already part of the short exact sequence 5.4(1), then the latter can be replaced by (1), which is split, with $\mathcal{B}'^{\bullet}$ isomorphic to $\mathcal{B}^{\bullet}$ and $C^{\bullet}_u$ q.i. to $C^{\bullet}$.

In this construction, u has been replaced by an injective direct morphism. It could similarly be replaced by a split surjection. In fact let $A'^{\bullet} = A^{\bullet} \oplus C^{\bullet}{}_{\mathcal{B}^{\bullet}}[-1]$ and define

(2) $$u'' : A'^{\bullet} = A^{\bullet} \oplus \mathcal{B}^{\bullet} \oplus \mathcal{B}^{\bullet}[-1] \to \mathcal{B}^{\bullet} ,$$

by

(3) $$u''(a,b,b') = u(a) - b .$$

Then we get a split exact sequence

(4) $$0 \to C^{\bullet}_u[-1] \to A'^{\bullet} \xrightarrow{u''} \mathcal{B}^{\bullet} \to 0,$$

which again leads to the previous long exact sequence (up to sign). If u is part of a short exact sequence

(5) $$0 \to C^{\bullet} \to A^{\bullet} \xrightarrow{u} \mathcal{B}^{\bullet} \to 0 ,$$

then we see that $C^{\bullet}_u[-1]$ is q.i. to $C^{\bullet}$, i.e. to ker u .

B. *Further properties of distinguished triangles* .

5.6 (Turning triangles). Let

(1) $$A^{\bullet} \xrightarrow{u} \mathcal{B}^{\bullet} \xrightarrow{v} C^{\bullet} \xrightarrow{w} A^{\bullet}[1] ,$$

be a distinguished triangle in K(X). Then

(2) $$\mathcal{B}^{\bullet} \xrightarrow{v} C^{\bullet} \xrightarrow{w} A^{\bullet}[1] \xrightarrow{-u[1]} \mathcal{B}^{\bullet}[1]$$

is a distinguished triangle.

We have to prove that $A^\bullet[1]$ is isomorphic to $C^\bullet_v$. For this we may assume that (1) is standard, i.e. $C^\bullet = C^\bullet_u$ and

(3) $$0 \to B^\bullet \xrightarrow{v} C^\bullet \to A^\bullet[1] \to 0$$

to be a short split exact sequence. But then $A^\bullet[1]$ is isomorphic to $C^\bullet_v$ by 5.4 (ii) and the diagram

$$\begin{array}{ccccccc} B^\bullet & \xrightarrow{v} & C^\bullet & \longrightarrow & C^\bullet_v & \longrightarrow & B^\bullet[1] \\ \downarrow{\scriptstyle Id} & & \downarrow{\scriptstyle Id} & & \downarrow{\scriptstyle m} & & \downarrow{\scriptstyle Id} \\ B^\bullet & \xrightarrow{v} & C^\bullet & \xrightarrow{w} & A^\bullet[1] & \xrightarrow{-u[1]} & B^\bullet[1] \end{array}$$

in which the top row is a standard triangle, is easily seen to commute up to homotopy.

5.7 Let

(1) $$A'^\bullet \xrightarrow{u'} B'^\bullet \xrightarrow{v'} C'^\bullet \xrightarrow{w'} A'^\bullet[1]$$

be a distinguished triangle. Then any (homotopy) commutative diagram

(2) $$\begin{array}{ccc} A^\bullet & \xrightarrow{u} & B^\bullet \\ \downarrow{\scriptstyle f} & & \downarrow{\scriptstyle g} \\ A'^\bullet & \xrightarrow{u'} & B'^\bullet \end{array}$$

extends to a morphism of 5.6(1) into 5.7(1) .

To see this, we may again assume (1) and 5.6(1) to be standard. By assumption (2) is homotopy commutative. There exists therefore a linear map $k : A^\bullet[1] \to B'^\bullet$ such that

(3) $$u' \circ f - g \circ u = dk + kd .$$

Define

(4) $$h : A^\cdot[1] \oplus B^\cdot \longrightarrow A'^\cdot[1] \oplus B'^\cdot ,$$

by

(5) $$h(a,b) = (fa, gb - ka) . \qquad (a \in A^{i+1}, b \in B^i ; i \in \mathbb{Z}) .$$

Then it is easily checked that h is a morphism and that the diagram

(6) $$\begin{array}{ccccc} B^\cdot & \longrightarrow & C_u^\cdot & \longrightarrow & A^\cdot[1] \\ \Big\downarrow g & & \Big\downarrow h & & \Big\downarrow f[1] \\ B'^\cdot & \longrightarrow & C_{u'}^\cdot & \longrightarrow & A'^\cdot[1] \end{array}$$

is commutative.

Note that since triangles can be turned (5.6), it also follows that a homotopy commutative diagram

$$\begin{array}{ccc} B^\cdot & \xrightarrow{v} & C^\cdot \\ \Big\downarrow & & \Big\downarrow \\ B'^\cdot & \xrightarrow{v'} & C'^\cdot \end{array} \qquad \text{or} \qquad \begin{array}{ccc} C^\cdot & \xrightarrow{w} & A^\cdot[1] \\ \Big\downarrow & & \Big\downarrow \\ C'^\cdot & \xrightarrow{w'} & A'^\cdot[1] \end{array}$$

extends to a morphism of triangles.

5.8 A diagram $A^\cdot \xrightarrow{f} B^\cdot \xleftarrow{u} C^\cdot$, where u is a q.i., can be completed to a homotopy commutative diagram

(1) $$\begin{array}{ccc} D^\cdot & \xrightarrow{g} & C^\cdot \\ w \Big\downarrow & & \Big\downarrow u \\ A^\cdot & \xrightarrow{f} & B^\cdot \end{array}$$

where w is a q.i.

To see this we consider the standard triangle

(2) $$C^\cdot \xrightarrow{u} B^\cdot \xrightarrow{v} C_u^\cdot \longrightarrow C^\cdot[1]$$

over u . Since u is a q.i. it follows from the long exact sequence that $H^{\cdot}C^{\cdot}_{u} = 0$. Let $h = v \circ f$. We have then a commutative diagram

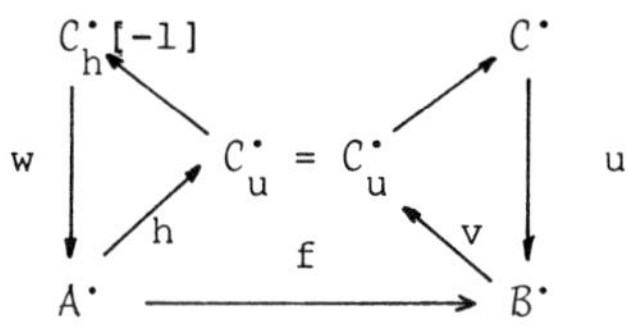

Note that w is a q.i., since $H^{\cdot}C^{\cdot}_{u} = 0$. By 5.8 , we can find $g : C^{\cdot}_{h} \to C^{\cdot}$ which completes (3) to a homotopy commutative diagram. We then take $\mathcal{D}^{\cdot} = C^{\cdot}_{h}[-1]$.

5.9 By reversing arrows, we see that similarly, a diagram $\mathcal{B}^{\cdot} \leftarrow A^{\cdot} \xrightarrow{u} C^{\cdot}$, where u is a q.i., can be completed to a homotopy commutative diagram

(1)

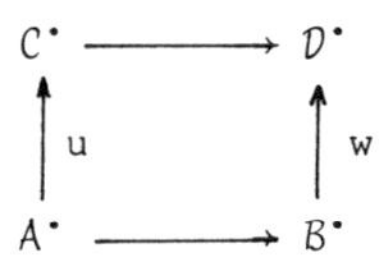

where w is a q.i.

5.10 All these constructions can be carried out with Sh(X) replaced by an arbitrary abelian category A . Instead of DGS(X) we consider the category C(A) of complexes of objects of A (with chain maps of degree 0 as morphisms), K(X) becomes K(A) and we get distinguished triangles in K(A) . If $F : Sh(X) \to A$ is a covariant additive functor, there are obvious extensions of F to $DGS(X) \to C(A)$ and $K(X) \to K(A)$, which we still denote F . It is clear that $F(\mathcal{S}^{\cdot}[1]) = F(\mathcal{S}^{\cdot})[1]$ and that F transforms distinguished triangles into distinguished triangles.

If $A^{\cdot},\mathcal{B}^{\cdot} \in DGS(X)$, $i \in \mathbb{Z}$; let $\mathrm{Hom}^{i}(A^{\cdot},\mathcal{B}^{\cdot})$ consist of all morphisms of graded sheaves $A^{\cdot} \to \mathcal{B}^{\cdot}$ which are homogenous of degree i . For $f \in \mathrm{Hom}^{i}(A^{\cdot},\mathcal{B}^{\cdot})$ let $d(f) = d_{\mathcal{B}^{\cdot}} \circ f + (-1)^{i+1} f \circ d_{A^{\cdot}}$. This defines a differential graded module $\mathrm{Hom}^{\cdot}(A^{\cdot},\mathcal{B}^{\cdot})$. Setting $\mathcal{H}om^{\cdot}(A^{\cdot},\mathcal{B}^{\cdot})(U) = \mathrm{Hom}^{\cdot}(A^{\cdot}|_{U},\mathcal{B}^{\cdot}|_{U})$ for U open in X, we get a DGS $\mathcal{H}om^{\cdot}(A^{\cdot},\mathcal{B}^{\cdot})$. We can view $\mathcal{H}om^{\cdot}$ as a functor from $DGS(X) \times DGS(X)$ to DGS(X) (contravariant in the first variable, covariant in the second) or from $K(X) \times K(X)$ to K(X) .

Let $\mathcal{D}^\cdot \in K(X)$ be fixed, and consider the functor $A^\cdot \longmapsto \mathcal{H}om^\cdot(\mathcal{D}^\cdot, A^\cdot)$. We have an obvious isomorphism $\mathcal{H}om^\cdot(\mathcal{D}^\cdot, A^\cdot[1]) = \mathcal{H}om^\cdot(\mathcal{D}^\cdot, A^\cdot)[1]$, and if the triangle

$$(1) \qquad A^\cdot \xrightarrow{u} B^\cdot \xrightarrow{v} C^\cdot \xrightarrow{w} A^\cdot[1]$$

is distinguished, then so is its image under $\mathcal{H}om^\cdot(\mathcal{D}^\cdot, \ )$

$$(2) \qquad \mathcal{H}om^\cdot(\mathcal{D}^\cdot, A^\cdot) \to \mathcal{H}om^\cdot(\mathcal{D}^\cdot, B^\cdot) \to \mathcal{H}om^\cdot(\mathcal{D}^\cdot, C^\cdot) \to \mathcal{H}om^\cdot(\mathcal{D}^\cdot, A^\cdot)[1]$$

It is enough to check this last point when (1) is standard. In this case $C^\cdot = C^\cdot_u$ and there is an obvious isomorphism

$$\mathcal{H}om^\cdot(\mathcal{D}^\cdot, C^\cdot_u) = C^\cdot_{\mathcal{H}om^\cdot(\mathcal{D}^\cdot, u)} .$$

Consider now the contravariant functor $A^\cdot \longmapsto \mathcal{H}om^\cdot(A^\cdot, \mathcal{D}^\cdot)$. The obvious isomorphism of graded sheaves $\mathcal{H}om^\cdot(A^\cdot[-1], \mathcal{D}^\cdot) \cong \mathcal{H}om^\cdot(A^\cdot, \mathcal{D}^\cdot)[1]$ is not an isomorphism of DGS. Let $\phi^i_{A^\cdot}: \mathcal{H}om^i(A^\cdot[-1], \mathcal{D}^\cdot) \to (\mathcal{H}om^\cdot(A^\cdot, \mathcal{D}^\cdot)[1])^i$ be multiplication by $(-1)^{i+1}$. Then $\phi_{A^\cdot} = (\phi^i_{A^\cdot})$ is an isomorphism of DGS. If the triangle (1) is distinguished, then so is

$$(3) \quad \mathcal{H}om^\cdot(C^\cdot, \mathcal{D}^\cdot) \to \mathcal{H}om^\cdot(B^\cdot, \mathcal{D}^\cdot) \to \mathcal{H}om^\cdot(A^\cdot, \mathcal{D}^\cdot) \to \mathcal{H}om^\cdot(C^\cdot, \mathcal{D}^\cdot)[1] ,$$

where the first two maps are induced by v and u, and the third one is $\phi_{C^\cdot} \circ \mathcal{H}om^\cdot(w[-1], \mathcal{D}^\cdot)$. This needs only to be checked in the case where

$$(4) \qquad B^\cdot \xrightarrow{v} C^\cdot \xrightarrow{w} A^\cdot[1] \xrightarrow{-u[1]} B^\cdot[1]$$

is standard, and the result follows from the existence of a suitable isomorphism $\mathcal{H}om^\cdot(C^\cdot_v[-1], \mathcal{D}^\cdot) \cong C^\cdot_{\mathcal{H}om^\cdot(v, \mathcal{D}^\cdot)}$.

In general a covariant (resp. contravariant) functor $F : K(X) \to K(A)$ is said to be exact if we are given a natural isomorphism $\phi_{A^\cdot} : F(A^\cdot[1]) \to F(A^\cdot)[1]$ (resp. $F(A^\cdot[-1]) \to F(A^\cdot)[1]$) and if the triangle $(F(A^\cdot), F(B^\cdot), F(C^\cdot), F(u), F(v), \phi_{A^\cdot} \circ F(w))$ (resp. $(F(C^\cdot), F(B^\cdot), F(A^\cdot), F(v), F(u), \phi_{C^\cdot} \circ F(w[-1])))$ is distinguished whenever the triangle (1) is distinguished.

5.11 We shall stop here this enumeration of properties of K(X). 5.3, 5.6, 5.7 almost prove that K(X), endowed with the set of distinguished triangles, is a "triangulated category" [7:p.20]. What is missing is mainly a proof of the "octahedral axiom", which we omit, since, as far as I can see, it is not needed in these Notes.

We note also that if $f : A^\bullet \to B^\bullet$ is a morphism and $A^\bullet, B^\bullet$ are bounded (or bounded above, or bounded below) then so is $C_f^\bullet$. Therefore if such boundedness conditions are satisfied by $A^\bullet, B^\bullet, C^\bullet$ is 5.8, 5.9, then the $D^\bullet$ constructed there also does so.

C. *The derived category of* Sh(X) .

5.12 The derived category D(X) of Sh(X) has the same objects as K(X), i.e. graded differential sheaves on X, but the morphisms are different.

A morphism from $A^\bullet$ to $B^\bullet$, $(A^\bullet, B^\bullet \in D(X))$ is an equivalence class of diagrams of morphisms in K(X) :

(1) $$A^\bullet \overset{\sim}{\leftarrow} C^\bullet \rightarrow B^\bullet$$

where $\overset{\sim}{\leftarrow}$ indicates a q.i.. This diagram is equivalent to $A^\bullet \overset{\sim}{\leftarrow} C'^\bullet \rightarrow B^\bullet$ if there exists $A^\bullet \overset{\sim}{\leftarrow} D^\bullet \to B^\bullet$ and morphisms $C^\bullet \leftarrow D^\bullet \to C'$ such that

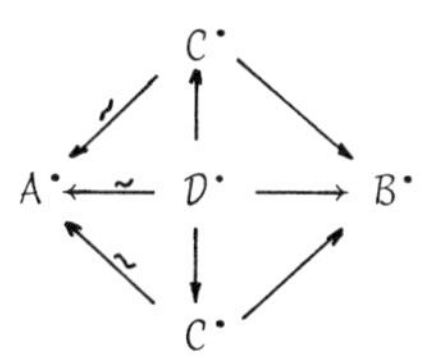

is commutative in K(X) (i.e. homotopy commutative). Another way to say this is to define

$$\mathrm{Mor}_{D(X)}(A^\bullet, B^\bullet) \text{ as } \varinjlim \mathrm{Mor}_{K(X)}(C^\bullet, B^\bullet) \text{ , where}$$

$\varinjlim$ is taken over morphisms which are parts of commutative diagrams

(2)

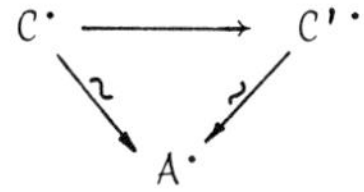

$A^\bullet$ and $B^\bullet$ are isomorphic in D(X) if there exists (1) where both maps

are q.i..

Similarly, one defines the derived category $D^b(X)$ of bounded complexes or the category $D^+(X)$ (resp. $D^-(X)$) of complexes which are bounded below (resp. above).

In case X is a pseudomanifold, another variant is to consider the subcategory of $D^b(X)$ consisting of bounded complexes which are cohomologically constructible with respect to some stratification, or with respect to a fixed stratification. In fact, the most important complexes considered in this seminar, namely those yielding intersection cohomology, are cohomologically constructible with respect to any stratification, as we saw in § 4 .

For the definition of D(X) to make sense, we need to know how to compose morphisms. Assume we have a diagram

(3)

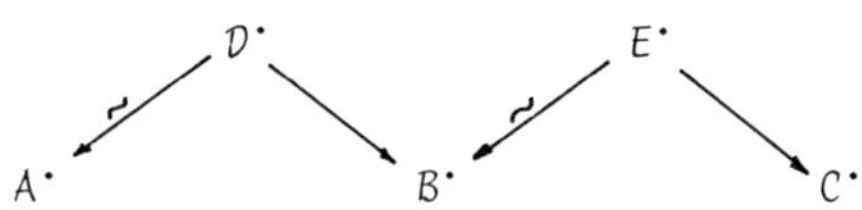

then, by 5.9 , we can complete it and get a homotopy commutative square :

(4)

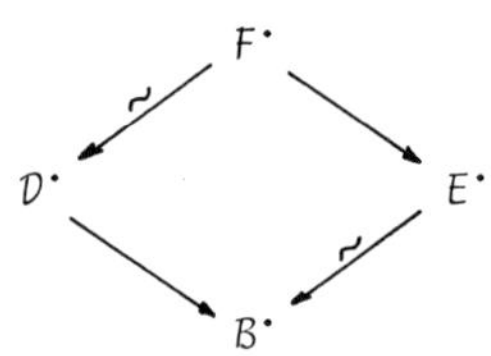

Then

(5) $$A^{\cdot} \xleftarrow{\simeq} F^{\cdot} \longrightarrow C^{\cdot}$$

is the composition of the given morphisms. Moreover, if all the morphisms in (3) are q.i. then so is $F^{\cdot} \to C^{\cdot}$ in (5). It should be checked of course that the equivalence class of (5) is independent from the choices made. This is left to the reader. For the proof, e.g. to verify that the choice of (4) is unimportant, it is useful to know the following fact [7:p.37] :

(6) *If* f,g : $A^\bullet \to B^\bullet$ *are morphisms in* K(X), *then the following conditions are equivalent :*

(i) *there exists a q.i.* s : $A'^\bullet \to A^\bullet$ *such that* f o s = g o s ;

(ii) *there exists a q.i.* t : $B^\bullet \to B'^\bullet$ *such that* t o f = t o g .

5.13 A *distinguished* triangle in D(X) is one which is isomorphic in D(X) to one in K(X) . It is clear that 5.6, 5.7 are valid in D(X) with minor changes in the formulation. Furthermore, the following properties of triangles in D(X) are easily deduced from 5.7 and the 5-lemma :

If the triangles $(A^\bullet,B^\bullet,C^\bullet,u,v,w)$ and $(A^\bullet,B^\bullet,C'^\bullet,u,v',w')$ are distinguished, then they are isomorphic.

If in a morphism of triangles, two of the maps are isomorphisms, then so is the third one.

Let $\mathfrak{X}$ be a topological stratification of X . If two of the objects in a distinguished triangle are $\mathfrak{X}$ - cc, then so is the third one.

5.14 As in 1.7, 1.8, let U be an open subset of X, Z be its complement and i : U → X, j : Z → X the inclusions. Then the exact sequence 1.8(6) now gives rise to a distinguished triangle in D(X)

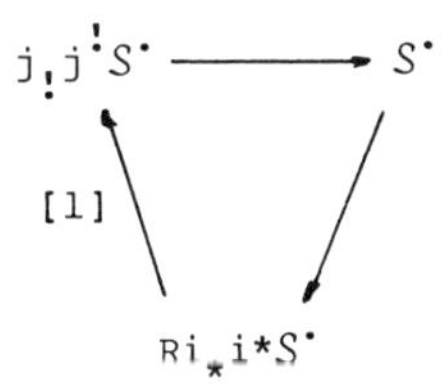

$(S^\bullet \in D(X))$

5.15 An element of $\mathrm{Mor}_{D(X)}(A^\bullet,B^\bullet)$ defines a unique morphism of $H^\bullet A^\bullet$ into $H^\bullet B^\bullet$, whence a natural map

$$\mathrm{Mor}_{D(X)}(A^\bullet,B^\bullet) \longrightarrow \mathrm{Mor}_{K(X)}(H^\bullet A^\bullet,H^\bullet B^\bullet) \ .$$

However this map is not an isomorphism in general [7:p.39].

As an example, start with an exact sequence

$$0 \longrightarrow A \xrightarrow{u} B \xrightarrow{v} C \longrightarrow 0 \qquad (1)$$

in Sh(X). Considering A,B,C as complexes, we get in D(X) a distinguished triangle

(2)

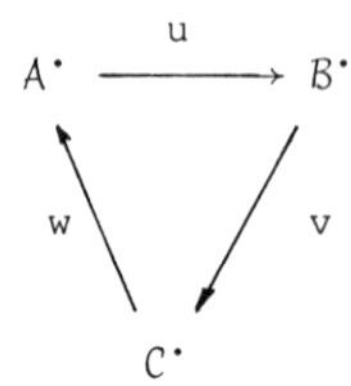

It is clear that $H^\cdot(w) : H^\cdot C^\cdot \to H^\cdot A^\cdot[1]$ is the zero map However, w need not be zero in general. Indeed, in the long exact sequence

$$\ldots \to H^i(X;\mathcal{A}) \to H^i(X;\mathcal{B}) \to H^i(X;\mathcal{C}) \to H^{i+1}(X;\mathcal{A}) \to \ldots$$

the connecting homomorphisms are induced by w . If $w = 0$, these connecting homomorphisms must also be zero, and this is not true in general.

5.16 If F is an additive functor from Sh(X) to an abelian category A , then the right derived functor $RF : D(X) \to D(A)$ is defined as in 1.5 by setting $RF(A^\cdot) = F(I^\cdot)$ for some injective right resolution $A^\cdot \to I^\cdot$. That this makes sense follows from the fact (implicitely used in 1.5 and standard for resolutions of a single sheaf) that all the injective resolutions of $A^\cdot$ are canonically isomorphic in K(X).

The key property to check this is the following. Let $I^\cdot, J^\cdot$ be injective complexes and let $u : I^\cdot \to J^\cdot$ be a q.i.. Then u is an isomorphism in K(X). To see this, notice that $C^\cdot = C^\cdot_u$ is an acyclic complex of injective sheaves. It follows then from 1.17 and our assumptions on X and R that for every n the sheaf $\ker d_n = \operatorname{coker} d_{n-1}$ is injective, hence that $C^\cdot$ is homotopic to 0 . Thus $Id_{C^\cdot} = dh + hd$ for some map $h : C^\cdot \to C^\cdot[-1]$. As $C^\cdot = I^\cdot[1] \oplus J^\cdot$, the application h gives maps

$$a : I^\cdot[1] \to I^\cdot \ , \ b : J^\cdot \to I^\cdot \ , \ c : J^\cdot \to J^\cdot[-1] \ .$$

It is readily checked that b is a chain map, and that a (resp. c) is a homotopy between b o u and $Id_{I^\cdot}$ (resp. $Id_{J^\cdot}$ and u o b).

From 5.9 we deduce then easily that if $f : A^\cdot \to B^\cdot$ is a morphism in K(X) and $s : A^\cdot \to C^\cdot$, $t : B^\cdot \to I^\cdot$ are resolutions with $I^\cdot$ injective, then there exists $g : C^\cdot \to I^\cdot$ such that g o s = t o f (in K(X)). We claim that g is unique in K(X). Indeed if $g' : C^\cdot \to I^\cdot$ has the same property, then g' o s = g o s. Hence by 5.12(6) there exists a q.i. $s' : I^\cdot \to J^\cdot$ such that s' o g' = s' o g . We can take $J^\cdot$ injective. Then s' is an isomorphism in K(X), and therefore g' = g .

This implies the required uniqueness of injective resolutions and shows also that RF is well defined on morphisms in D(X).

It is clear that RF transforms distinguished triangles into distinguished triangles.

For example if $j : Z \to X$ is a closed immersion the functor $j^{!}$ is a functor from D(X) to D(Z).

We can define in a similar way $RF : D(X) \to D(A)$ when F is an exact functor from K(X) to K(A) (5.10).

To define $R\mathcal{H}om^{\bullet}$, we consider $\mathcal{H}om^{\bullet}(A^{\bullet},B^{\bullet})$ as a functor of $B^{\bullet}$ (with $A^{\bullet}$ fixed) and we take its right derived functor. We shall see in 5.17 that for a fixed $B^{\bullet}$, the functor $A^{\bullet} \longmapsto R\mathcal{H}om^{\bullet}(A^{\bullet},B^{\bullet})$ transforms q.i. into q.i., hence defines a functor from D(X) into itself.

The functor $RHom^{\bullet}$ is defined in a similar way. For $A^{\bullet},B^{\bullet} \in D(X)$ we define also

$$\mathrm{Ext}^{\bullet}(A^{\bullet},B^{\bullet}) = H^{\bullet}(RHom^{\bullet}(A^{\bullet},B^{\bullet})) .$$

5.17 For any $A^{\bullet},B^{\bullet} \in K(X)$, there is a natural homomorphism

(1) $$\mathrm{Mor}_{K(X)}(A^{\bullet},B^{\bullet}) \longrightarrow \mathrm{Mor}_{D(X)}(A^{\bullet},B^{\bullet}) .$$

It follows from the discussion in 5.16 that (1) is an isomorphism if $B^{\bullet}$ is injective. This gives another description of the morphisms in D(X).

We can also recover $\mathrm{Mor}_{D(X)}(A^{\bullet},B^{\bullet})$ from $RHom^{\bullet}(A^{\bullet},B^{\bullet})$ or $R\mathcal{H}om^{\bullet}(A^{\bullet},B^{\bullet})$. Let $I^{\bullet}$ be an injective resolution of $B^{\bullet}$. It follows immediately from the definition of the differential on $Hom^{\bullet}(A^{\bullet},I^{\bullet})$ (5.10) that $H^{i}(Hom^{\bullet}(A^{\bullet},I^{\bullet})) = \mathrm{Mor}_{K(X)}(A^{\bullet},I^{\bullet}[i])$, and this is $\mathrm{Mor}_{D(X)}(A^{\bullet},I^{\bullet}[i]) = \mathrm{Mor}_{D(X)}(A^{\bullet},B^{\bullet}[i])$ by (1) . Thus

(2) $$\mathrm{Ext}^{i}(A^{\bullet},B^{\bullet}) = \mathrm{Mor}_{D(X)}(A^{\bullet},B^{\bullet}[i]) .$$

As $I^{\bullet}$ is injective, $\mathcal{H}om^{\bullet}(A^{\bullet},I^{\bullet})$ is a complex of flabby sheaves [5:II.7.3.2], hence can be used to compute hypercohomology. Therefore

$$\mathbb{H}^\bullet(X;R\mathcal{H}om^\bullet(A^\bullet,B^\bullet)) = H^\bullet(\Gamma(X;\mathcal{H}om^\bullet(A^\bullet,I^\bullet))$$

$$= H^\bullet(\mathrm{Hom}^\bullet(A^\bullet,I^\bullet)) = \mathrm{Ext}^\bullet(A^\bullet,B^\bullet)$$

and (2) gives also

$$\text{(3)} \qquad \mathbb{H}^i(X;R\mathcal{H}om^\bullet(A^\bullet,B^\bullet)) = \mathrm{Mor}_{D(X)}(A^\bullet,B^\bullet[i]) .$$

The same holds over open subsets of X . The limit over the open neighborhoods of $x \in X$ of the left hand side of (3) gives $H^\bullet(R\mathcal{H}om^\bullet(A^\bullet,B^\bullet)_x)$. But the right hand side of (3) depends on $A^\bullet$ only up to q.i.. Thus for a fixed $B^\bullet$, the functor $A^\bullet \longmapsto R\mathcal{H}om^\bullet(A^\bullet,B^\bullet)$ transforms q.i. into q.i., as claimed in 5.16 .

5.18 It is often convenient to use resolutions by F - acyclic objects to compute RF. For example let us check that we can indeed use flabby resolutions to compute $R\Gamma(X;\ )$ (and hence hypercohomology). In view of 5.9, we must check that if $u : A^\bullet \to B^\bullet$ is a q.i. and $A^\bullet,B^\bullet$ are flabby, then the morphism $\Gamma(X;A^\bullet) \to \Gamma(X;B^\bullet)$ induced by u is still a q.i.; or equivalently, that $\Gamma(X;C_u^\bullet)$ is an acyclic complex of R-modules. Let d be the differential of $C_u^\bullet$. Since $C_u^\bullet$ is an acyclic complex of flabby sheaves, it follows from 1.17(i) that for every $n \in \mathbb{Z}$ the sheaf $\ker d_n = \mathrm{Im}\, d_{n-1}$ is also flabby.The acyclicity of $\Gamma(X;C_u^\bullet)$ is then clear.

More generally let A,B be abelian categories and let $F : A \to B$ be an additive functor. If A has enough F - acyclic objects (in some precise technical sense) and for some $n \in \mathbb{N}$ every object has an F - acyclic resolution of length $\leqq n$, then the argument above can be used to define RF even if A has not enough injectives, or if the injective dimension of A is not finite. We shall encounter a similar situation with the definition of the left derived functor of $\otimes$ in D(X) (6.2) in a case where there are not enough projectives, but where an ad hoc argument will show that the definition makes sense.

In some cases RF can also be defined only on some subcategories of D(X), e.g. on $D^+(X)$.

Remark . 5.16, which expands an earlier version, and 5.17, 5.18 were written up by N. Spaltenstein.

§ 6 FLAT AND c-SOFT SHEAVES

We review or prove here some facts about flat or c-soft sheaves, and recall the definition of the left derived functors of the tensor product. In this paragraph and the next one, we lean heavily on B.Iversen's Notes [8]. As usual, the underlying space X is locally compact, of finite cohomological dimension n over R and R has finite cohomological dimension d. We let Mod(R) denote the category of R-modules.

6.0 *Flat modules*. We recall that a R-module is *flat* if it satisfies the following equivalent conditions :

(i) For any monomorphism $i : B \to C$ in Mod(R), the morphism $i \otimes \mathrm{Id.}: B \otimes A \to C \otimes A$ is injective.

(ii) The functor $\otimes A$ is exact.

(iii) $\mathrm{Tor}_i(A,B) = 0$ for any $B \in \mathrm{Mod}(R)$ and $i \geqq 1$.

(iv) $\mathrm{Tor}_1(A,B) = 0$ for any $B \in \mathrm{Mod}(R)$.

(see e.g. [3], Prop.3,p.8 and Théor.2,p.74).

The functors Tor_i are usually defined by means of left free or projective resolutions but flat ones can also be used [3:Théor.1,p.100]. The equality dim R = d implies

(1) $$\mathrm{Tor}_i(A,B) = 0 \text{ for } A,B \in \mathrm{Mod}(R) \text{ and } i > d .$$

[3: lemme 1,p.134] . Let

(2) $$0 \to A \to B_{k-1} \to \dots \to B_0 \to C \to 0 ,$$

be an exact sequence in Mod(R), where the B_i's are flat. The standard shift argument by means of long exact sequences implies

(3) $$\mathrm{Tor}_i(A,D) = \mathrm{Tor}_{i+k}(C,D) \qquad (i \geqq 1 , D \in \mathrm{Mod}(R)).$$

It follows then from the above that A is flat if $k \geqq d$. In particular, every R-module C has a flat left resolution of length $\leqq d$. More generally, it also follows from (3) that if C has a left flat resolution of length e and $k \geqq e$, then A is flat.

6.1 A sheaf $\mathcal{A}$ on X is *flat* (or R-flat) if, given a monomorphism $i : \mathcal{B} \to \mathcal{C}$, the morphism $i \otimes 1 : \mathcal{B} \otimes \mathcal{A} \to \mathcal{C} \otimes \mathcal{A}$ is also injective. The sheaf $\mathcal{A}$ is flat if and only if $\mathcal{A}_x$ is a flat R-module for all $x \in X$. A complex of sheaves is flat if it consists of flat sheaves.

Given a sheaf $\mathcal{A}$, there always exists a surjective morphism $\mathcal{P} \to \mathcal{A}$, where $\mathcal{P}$ is a direct sum of sheaves R_U (U open in X) extended by zero. (1.11(a)). Those are obviously flat, hence so is $\mathcal{P}$. From (6.0) we see then:

(1) *Every* $\mathcal{A} \in Sh(X)$ *has a left flat resolution of length* $\leq d = \dim R$

More generally, given $\mathcal{A}^\bullet \in DGS(X)$, there exists a left flat resolution $\mathcal{P}^\bullet \to \mathcal{A}^\bullet$. Such a resolution can be constructed as follows. For each i, let $f_i : \mathcal{F}^i \to \mathcal{A}^i$ be a surjective morphism of sheaves with $\mathcal{F}^i$ flat and $\mathcal{F}^i = 0$ if $\mathcal{A}^i = 0$. Let $\mathcal{Q}^\bullet$ be the complex defined by $\mathcal{Q}^i = \mathcal{F}^i \oplus \mathcal{F}^{i-1}$, $d(x,y) = (0,x)$ $(x \in \mathcal{F}^i,\ y \in \mathcal{F}^{i-1})$ and define $v : \mathcal{Q}^\bullet \to \mathcal{A}^\bullet$ by $v(x,y) = f_i(x) + df_{i-1}(y)$. Then $\mathcal{Q}^\bullet$ is flat and v is a surjective chain map. We have a short exact sequence

(2) $$0 \to \operatorname{Ker} v \xrightarrow{u} \mathcal{Q}^\bullet \xrightarrow{v} \mathcal{A}^\bullet \to 0 ,$$

and by 5.4(i) a q.i. $C_u^\bullet \to \mathcal{A}^\bullet$. Let $\mathcal{P}_1^\bullet = C_u^\bullet$. Iterating this operation, we get a sequence of q.i.'s

(3) $$\cdots \to \mathcal{P}_k^\bullet \to \mathcal{P}_{k-1}^\bullet \to \cdots \to \mathcal{P}_1^\bullet \to \mathcal{A}^\bullet ,$$

and it is therefore sufficient to check that $\mathcal{P}_d^\bullet$ is flat.

If for some $e \in \mathbb{N}$, each $\mathcal{A}^i$ in (2) has a left flat resolution of length $\leq e$, then, by 6.0, each $(\ker v)^i$ has a left flat resolution of length $\leq \max(e-1,0)$, and so does each $C_u^i = (\ker v)^{i+1} \oplus \mathcal{Q}^i$. Using (1) and induction on k, we find that in general $\mathcal{P}_k^i$ has a left flat resolution of length $\leq \max(d-k,0)$. In particular $\mathcal{P}_d^\bullet$ is flat and we can take $\mathcal{P}^\bullet = \mathcal{P}_d^\bullet$.

It is clear that if $\mathcal{A}^\bullet$ is bounded (resp. bounded above, resp. bounded below), then so is the flat resolution $\mathcal{P}^\bullet$ given by this construction.

Remark: Originally, I had used a construction suggested in [8], which applies to bounded above complexes. The previous argument is due to N. Spaltenstein.

6.2 Let $\mathcal{P}^\cdot \in DGS^-(X)$ be flat. Let $\mathcal{Q}^\cdot, \mathcal{B}^\cdot \in DGS(X)$ and $\mathcal{Q}^\cdot \to \mathcal{B}^\cdot$ be a q.i. We claim that the induced homomorphism $m : \mathcal{P}^\cdot \otimes \mathcal{Q}^\cdot \to \mathcal{P}^\cdot \otimes \mathcal{B}^\cdot$ is also a q.i.. We note first that we can assume $\mathcal{Q}^\cdot$ and $\mathcal{B}^\cdot$ to be bounded above. In fact, any $\mathcal{S}^\cdot \in DGS(X)$ is the inductive limit of its truncations $\tau_{\leq k}\mathcal{S}^\cdot$. Since $\otimes$ and taking cohomology commute with inductive limits, it is enough to prove our assertion with $\mathcal{Q}^\cdot$ and $\mathcal{B}^\cdot$ replaced by their truncations at a given level. Consider now the spectral sequence (E_r) and (E'_r) associated to the filtration of $\mathcal{P}^\cdot \otimes \mathcal{Q}^\cdot$ and $\mathcal{P}^\cdot \otimes \mathcal{B}^\cdot$ by the degree in $\mathcal{P}^\cdot$. Since $\mathcal{P}^\cdot$ is flat, we have

$$E_1 = \mathcal{P}^\cdot \otimes \mathcal{H}^\cdot\mathcal{Q}^\cdot \quad , \quad E'_1 = \mathcal{P}^\cdot \otimes \mathcal{H}^\cdot\mathcal{B}^\cdot \ ,$$

hence the map $(E_r) \to (E'_r)$ induced by m is an isomorphism at the E_1-level. In view of our boundedness assumptions,these spectral sequences converge and our claim follows.

Let now $\mathcal{A}^\cdot, \mathcal{B}^\cdot \in DGS^-(X)$. Choose, as we may by 6.1, flat left resolutions $\mathcal{P}^\cdot, \mathcal{Q}^\cdot$ of $\mathcal{A}^\cdot$ and $\mathcal{B}^\cdot$ respectively which are bounded above. The result just proved implies that we have q.i.

$$(2) \qquad \mathcal{P}^\cdot \otimes \mathcal{B}^\cdot = \mathcal{P}^\cdot \otimes \mathcal{Q}^\cdot = \mathcal{A}^\cdot \otimes \mathcal{Q}^\cdot .$$

We then define the left derived functor $\mathcal{A}^\cdot \overset{L}{\otimes} \mathcal{B}^\cdot$ in $D^-(X)$ by

$$(3) \qquad \mathcal{A}^\cdot \overset{L}{\otimes} \mathcal{B}^\cdot = \mathcal{P}^\cdot \otimes \mathcal{B}^\cdot = \mathcal{A}^\cdot \otimes \mathcal{Q}^\cdot = \mathcal{P}^\cdot \otimes \mathcal{Q}^\cdot .$$

The ith left derived functor is then

$$(4) \qquad \mathcal{T}or_i(\mathcal{A}^\cdot, \mathcal{B}^\cdot) = \mathcal{H}^{-i}(\mathcal{A}^\cdot \overset{L}{\otimes} \mathcal{B}^\cdot) \qquad (i \in \mathbb{Z}) .$$

If $\mathcal{A}^\cdot, \mathcal{B}^\cdot \in DGS^b(X)$, then we may also take $\mathcal{P}^\cdot, \mathcal{Q}^\cdot$ bounded and $\mathcal{A}^\cdot \overset{L}{\otimes} \mathcal{B}^\cdot$ is defined in $D^b(X)$.

It is also possible to define $\mathcal{A}^\cdot \overset{L}{\otimes} \mathcal{B}^\cdot$ in D(X) without boundedness conditions (see 6.9), but we shall not need this.

6.3 We recall that a sheaf $\mathcal{A}$ is c-soft if any continuous section over a compact subset is the restriction of a continuous section on X (this is called soft in [8]). If $\mathcal{A}$ is c-soft, then $H^i_c(X;\mathcal{A}) = 0$ for $i \geq 1$. It follows from the cohomology sequence with compact supports that $\mathcal{A}$ is c-soft if and only if $H^1_c(U;\mathcal{A}) = 0$ for all open $U \subset X$.

Any injective or flabby sheaf is c-soft . If a sheaf on the open subset $U \subset X$ is c-soft, then its extension by zero is c-soft on X. The restriction of a c-soft sheaf to a closed subspace is c-soft.

6.4 Consider an exact sequence of sheaves

(1) $$0 \to \mathcal{A} \to \mathcal{F}_{k-1} \to \dots \to \mathcal{F}_0 \to \mathcal{B} \to 0$$

where the $\mathcal{F}_i$ are c-soft. Since the $\mathcal{F}_i$ are acyclic for cohomology with compact supports it follows by standard arguments that

(2) $$H_c^{i+k}(U;\mathcal{A}) = H_c^i(U;\mathcal{B}) \qquad (U \text{ open in } X;\ i \geqq 1) .$$

If now $k \geqq n$, then $H_c^{k+i}(U;\mathcal{A}) = 0$, hence $H_c^i(U;\mathcal{B}) = 0$ and consequently $\mathcal{B}$ is c-soft (6.3). This also shows that if $k \geqq 0$ and $\mathcal{A}$ is also c-soft, then $\mathcal{B}$ is c-soft.

6.5 PROPOSITION . *Let* $\mathcal{S},\mathcal{A} \in \mathrm{Sh}(X)$. *Assume that* $\mathcal{S}$ *is c-soft and that either* $\mathcal{A}$ *or* $\mathcal{S}$ *is flat. Then* $\mathcal{A} \otimes \mathcal{S}$ *is c-soft.*

Proof. There is an exact sequence

(1) $$0 \to \mathcal{P}_n \to \mathcal{P}_{n-1} \to \dots \to \mathcal{P}_0 \to \mathcal{A} \to 0 ,$$

where $\mathcal{P}_j$ is a direct sum of sheaves R_U $(0 \leqq j \leqq n-1)$ (6.1), hence is flat. The following sequence

(2) $$0 \to \mathcal{P}_n \otimes \mathcal{S} \to \mathcal{P}_{n-1} \otimes \mathcal{S} \to \dots \to \mathcal{P}_0 \otimes \mathcal{S} \to \mathcal{A} \otimes \mathcal{S} \to 0$$

is also exact because either $\mathcal{S}$ is flat or, if $\mathcal{A}$ is flat, all the terms in (1) are flat. For any open $U \subset X$, the sheaf $R_U \otimes \mathcal{S}$ is the extension of $\mathcal{S}|U$ by zero, hence is c-soft. Therefore $\mathcal{P}_i \otimes \mathcal{S}$ is c-soft $(0 \leqq i \leqq n-1)$. Then $\mathcal{A} \otimes \mathcal{S}$ is c-soft by 6.4 .

Since c-soft sheaves are acyclic for cohomology with compact supports, 6.5 implies :

6.6 COROLLARY *Let* S *be flat and c-soft. Then the functor* $A \mapsto \Gamma_c(A \otimes S)$ *is exact.*

6.7 PROPOSITION . (i) *Let* F *be a flat sheaf. Then* F *has a c-soft flat resolution* $F \to K^{\cdot}$ *, where* $K^i = 0$ *for* $i > n$.

(ii) *Let* $R_X \to S^{\cdot}$ *be a c-soft flat resolution of* R_X *, such that* $S^i = 0$ *for* $i > n$. *Then, for any sheaf* A, *the canonical morphism* $A \to A \otimes S^{\cdot}$ *is a c-soft resolution of* A . *In particular,* A *has a c-soft resolution which vanishes in degrees* $i > n$.

Proof : (i) We consider first the canonical Godement resolution

(1) $$0 \to F \to J^0 \xrightarrow{d_0} J^1 \xrightarrow{d_1} \cdots \longrightarrow J^{n-1} \xrightarrow{d_{n-1}} \cdots$$

The J^i's, being flabby, are c-soft. We claim that J^i and Im d_i are flat. It suffices to prove this for $j = 0$. For U open in X, we have $J^0(U) = \Pi_{x\in U} F_x$, hence $J^0(U)$ is flat, since R is noetherian. The R-module J^0_x , being an inductive limit of flat R- modules, is then also R-flat. Moreover J^0_x is the direct sum of F_x and of an R-submodule K_x (the sections with value 0 at x). Then K_x is flat. Since it is isomorphic to $(\mathrm{Im}\ d_0)_x$, the latter is flat. This proves our assertion. We then take as a resolution of F :

(2) $$0 \to F \to S^0 \to \cdots \to S^n \to 0,$$

where

(3) $$S^i = J^i \quad (0 \leqq i \leqq n-1) \ , \ S^n = \mathrm{Im}\ d_{n-1} \ ,$$

and observe that S^n is c-soft by 6.4 .

(ii) Let $R_X \to S^{\cdot}$ be as in (ii). Then

$$A = A \otimes R_X \to A \otimes S^{\cdot}$$

is a resolution, since the S^i's are flat, and is c-soft by 6.5 .

6.8 COROLLARY . *If* $A^{\cdot}$ *is a bounded complex of sheaves, then* $A^{\cdot} \to A^{\cdot} \otimes S^{\cdot}$ *is a bounded c-soft resolution of* $A^{\cdot}$.

6.9 (N. Spaltenstein): Let $\mathcal{P}^\bullet \to \mathcal{A}^\bullet$, $\mathcal{Q}^\bullet \to \mathcal{B}^\bullet$ be flat resolutions of $\mathcal{A}^\bullet, \mathcal{B}^\bullet \in K(X)$, with $\mathcal{A}^\bullet, \mathcal{B}^\bullet$ not necessarily bounded above. We check here that, as in 6.2 , we have quasi-isomorphisms

(1) $$\mathcal{P}^\bullet \otimes \mathcal{Q}^\bullet = \mathcal{P}^\bullet \otimes \mathcal{B}^\bullet = \mathcal{A}^\bullet \otimes \mathcal{Q}^\bullet ,$$

so that $\mathcal{A}^\bullet \overset{L}{\otimes} \mathcal{B}^\bullet$ is well-defined and can be computed by any one of the expressions in (1) .

Using cones and the symmetry between $\mathcal{A}^\bullet$ and $\mathcal{B}^\bullet$, we see easily that it suffices to prove the following .

Let $\mathcal{A}^\bullet$ be flat :

(a) If $\mathcal{A}^\bullet$ is acyclic, then so is $\mathcal{A}^\bullet \otimes \mathcal{B}^\bullet$.

(b) If $\mathcal{B}^\bullet$ is acyclic, then so is $\mathcal{A}^\bullet \otimes \mathcal{B}^\bullet$.

As this needs only to be checked on stalks, we may (and do) assume that we are dealing with complexes of R-modules.

Let d be the differential of $A^\bullet$. Assume that $A^\bullet$ is not only flat, but that

(2) $$\text{Ker } d_i \text{ and Im } d_i \text{ are flat } (i \in \mathbb{Z}) .$$

Then, by [3,p.76-77], there is a Künneth rule for $H^\bullet(A^\bullet \otimes B^\bullet)$. In particular $H^\bullet(A^\bullet \otimes B^\bullet) = 0$ if $A^\bullet$ or $B^\bullet$ is acyclic.

In view of 6.0 , it is easily checked that (2) holds if $A^\bullet$ and $H^\bullet(A^\bullet)$ are both flat. This proves (a) and a special case of (b).
To prove (b) in general, let $e(A^\bullet)$ be the smallest integer e such that every module $H^i(A^\bullet)$ has a flat resolution of length $\leq e$. By 6.0, $e(A^\bullet)$ is finite. We use induction on $e(A^\bullet)$. We know already that (b) holds if $e(A^\bullet) = 0$. For the induction step, it is sufficient to show that if $e(A^\bullet) \geq 1$, then there exists a morphism $f : F^\bullet \to A^\bullet$ with $F^\bullet$ flat, $e(F^\bullet) = 0$, $e(C_f^\bullet) = e(A^\bullet) - 1$. Such a morphism can be constructed as follows : for each i , let $F^i \to \text{Ker } d_i$ be a surjective map with F^i flat. Give $F^\bullet$ the zero differential .
Since $H^\bullet(F^\bullet) = F^\bullet \to H^\bullet(A^\bullet)$ is surjective, we have a short exact sequence

$$0 \to H^\bullet(C_f^\bullet)[-1] \to F^\bullet \to H^\bullet(A^\bullet) \to 0 .$$

Then 6.0 shows that $f : F^\bullet \to A^\bullet$ has the required properties.

§ 7 THE DUAL OF A COMPLEX OF SHEAVES. VERDIER DUALITY

We fix an injective resolution $I^\bullet$ of R, of length equal to the dimension d of R, and a c-soft flat resolution $\mathcal{K}^\bullet$ of R_X, of length n = dim X. For U open in X, and $\mathcal{A}$ a sheaf on X, we let $\mathcal{A}_U$ denote the restriction of $\mathcal{A}$ to U. In particular, $\mathcal{K}_U^\bullet$ is a c-soft flat resolution of R_U. Sometimes, we shall also denote in this way the extension by zero to X of the restriction of $\mathcal{A}$ to U, i.e. make no notational distinction between $\mathcal{A}_U$ and $j_!\mathcal{A}_U$.

A. *The dualizing sheaf and homology*.

7.1 We consider the presheaf

(1) $$U \longmapsto \mathrm{Hom}^\bullet(\Gamma_c(\mathcal{K}_U^\bullet), I^\bullet).$$

It is a sheaf ([2], [4:V,1.9]), to be called the *dualizing sheaf* on X and to be denoted by $\mathcal{D}_X^\bullet$. We have

(2) $$\mathcal{D}_X^i(U) = \oplus_j \mathrm{Hom}(\Gamma_c(\mathcal{K}_U^j), I^{j+i}), \quad (i \in \mathbb{Z})$$

hence, with our choice of $\mathcal{K}^\bullet$ and $I^\bullet$, it can be non-zero only for $i \in [-n,d]$. [The same definition with another injective resolution of R or another c-soft resolution of R_X, (not necessarily flat), leads to a q.i. complex of sheaves, hence $\mathcal{D}_X^\bullet$ is well defined in $D^b(X)$]. For $I^\bullet$ injective, $\mathcal{D}_X^\bullet$ is obviously flabby. We shall see that, with $\mathcal{K}^\bullet$ flat, it is in fact injective (7.6). It follows therefore from the definitions that we have

(3) $$\mathbf{H}^\bullet(U;\mathcal{D}_X^\bullet) = \mathrm{Ext}^\bullet(\Gamma_c(\mathcal{K}_U^\bullet), R).$$

We recall that the right hand side is the abutment of a spectral sequence in which

(4) $$E_2^{p,q} = \mathrm{Ext}^p(H_c^{-q}(U;R), R) \quad (p,q \in \mathbf{Z})$$

If d = 1 , in particular if R is a Dedekind ring, then (3) yields an exact sequence

(5) $$0 \to \mathrm{Ext}(\mathbf{H}_c^{i+1}(U;R),R) \to \mathbf{H}^{-i}(X;\mathcal{D}_X^{\bullet}) \to \mathrm{Hom}(\mathbf{H}_c^{i}(U;R) \to 0 ,$$

which becomes, when R is a field :

(6) $$\mathbf{H}^{i}(U;\mathcal{D}_X^{\bullet}) = \mathrm{Hom}(H_c^{-i}(U;R),R) .$$

Assume that X satisfies the first axiom of countability and that R_X is cc (§ 3). Then, taking the limit over neighborhoods U of x ,we get from (3)

(7) $$(\mathcal{D}_X^{\bullet})_x = R\ \mathrm{Hom}^{\bullet}(f_x^{!}R_X,R)$$

in particular

(8) $$H^{\bullet}(\mathcal{D}_X^{\bullet})_x = \mathrm{Ext}^{\bullet}(f_x^{!}K^{\bullet},R) ,$$

where the right hand side is the abutment of a spectral sequence in which

(9) $$E_2^{p,q} = \mathrm{Ext}^p(H^{-q}(f_x^{!}R_X),R) \qquad (p,q \in \mathbb{Z}) .$$

7.2 By definition in [2] , $\mathbf{H}^{-i}(X,\mathcal{D}_X^{\bullet}) = H^{-i}(\Gamma(\mathcal{D}_X^{\bullet}))$ is the i-th *homology group* of X with coefficients in R and arbitrary closed supports. A priori, this group may be non-zero for $i \in [-d,n]$. One would of course like it to be zero for $i < 0$. This is proved in [9] when R is a principal ideal domain (which implies $d \leqq 1$). I do not know what is the situation for more general rings. If X has reasonable local properties e.g. if it is locally constructible, the homology is indeed zero in degrees $i < 0$.

The sheaf $\mathcal{C}.$ defined by $\mathcal{C}_i = \mathcal{D}_X^{-i}$ $(i \in \mathbb{Z})$ is the fundamental homology sheaf. If X has a PL structure, it is q.i. to the homology sheaf constructed in I,II .

7.3 If X is a manifold, then

(1) $$H_i C_\bullet = 0 \quad (i \neq n)$$

(2) $$H_n C_\bullet = \mathcal{O} \quad (\mathcal{O}\text{: orientation sheaf}) .$$

Therefore $\mathcal{D}^\bullet_X[-n]$ is a resolution of $\mathcal{O}$, (which is injective by 7.7).

We recall that a cohomology n-manifold over R can be defined as a finite dimensional space where R_X is cc , in which (1) is valid and $H_n C_\bullet$ is locally isomorphic to R_X . However we shall not need this notion in this seminar.

B. *The dual of a complex of sheaves* .

In this section, our main goal is to prove the equality in $D^b(X)$ of the dual of a complex of sheaves $A^\bullet$, as constructed in [2] , and of the Verdier dual of $A^\bullet$ [10]. This theorem can be viewed as a special case of the Verdier duality, which will be discussed in general in the next section.

The main point in the proof is an isomorphism at the sheaf level. We discuss this first.

7.4 We fix an R-module N and a flat c-soft sheaf K . For a sheaf A , we consider the presheaf

(1) $$E(A) : U \longmapsto \mathrm{Hom}(\Gamma_c((A \otimes K)_U), N) \quad (U \text{ open in } X)$$

It is a sheaf, as easily seen [2;8] .

7.5 PROPOSITION . *There is a natural isomorphism of sheaves*

(1) $$\mu_A : E(A) = \mathcal{H}om(A, E(R_X)) .$$

If N *is injective, the functor* $A \longmapsto E(A)$ *is exact and* $E(R_X)$ *is injective.*

Let $V \subset U$ be open in X . We have natural homomorphisms

(2) $$A(V) \otimes \Gamma_c(K_V) \to \Gamma_c((A \otimes K)_V) \to \Gamma_c((A \otimes K)_U) ,$$

whence

$$(3) \qquad \mathrm{Hom}(\Gamma_c((\mathcal{A} \otimes \mathcal{K})_U),N) \to \mathrm{Hom}(\mathcal{A}(V) \otimes \Gamma_c(\mathcal{K}_V),N) .$$

But the right hand side is equal to

$$(4) \qquad \mathrm{Hom}(\mathcal{A}(V),\mathrm{Hom}(\Gamma_c(\mathcal{K}_V),N)) = \mathrm{Hom}(\mathcal{A}(V),\mathcal{E}(R_X)(V)) ,$$

whence a homomorphism

$$(5) \qquad \mu_V : \mathcal{E}(\mathcal{A})(U) \to \mathrm{Hom}(\mathcal{A}(V),\mathcal{E}(R_X)(V)) .$$

For $V' \subset V$ open and $\sigma \in \mathcal{E}(\mathcal{A})(U)$, the following diagram is commutative:

$$\begin{array}{ccc} \mathcal{A}(V) & \xrightarrow{\ \mu_V(\sigma)\ } & \mathcal{E}(R_X)(V) \\ \downarrow & & \downarrow \\ \mathcal{A}(V') & \xrightarrow{\ \mu_{V'}(\sigma)\ } & \mathcal{E}(R_X)(V') \end{array}$$

Hence we get

$$(6) \qquad \mu : \mathcal{E}(\mathcal{A})(U) \longrightarrow \mathcal{H}om(\mathcal{A},\mathcal{E}(R_X))(U) .$$

This again commutes with restriction to $U' \subset U$, and yields a sheaf homomorphism

$$(7) \qquad \mu_{\mathcal{A}} : \mathcal{E}(\mathcal{A}) \longrightarrow \mathcal{H}om(\mathcal{A},\mathcal{E}(R_X)) .$$

This is the natural homomorphism in (1). To prove it is an isomorphism, we have to show that

$$(8) \qquad \mathcal{E}(\mathcal{A})(U) = \mathrm{Hom}(\mathcal{A}_U,\mathcal{E}(R_X)_U) , \qquad (U \text{ open in } X) .$$

Of course, $\mathcal{E}(R_X)_U = \mathcal{E}(R_U)$ on U. So we have to prove

$$(9) \qquad \mathcal{E}(\mathcal{A})(U) = \mathrm{Hom}(\mathcal{A}_U,\mathcal{E}(R_U)) , \qquad (U \text{ open in } X) .$$

Assume first that $A = R_V$ for V open in U . We have $R_V \otimes K = K_V$, therefore $E(R_V)(U)$ is equal to $E(R_V)(V)$. But

$$\text{(10)} \qquad \mathrm{Hom}(R_V, E(R_U)) = E(R_U)(V) = E(R_V)(V) .$$

This proves (8) in that case, hence also when A is a direct sum of sheaves R_V .
As recalled in 6.1 , there is an exact sequence

$$\text{(11)} \qquad Q \to P \to A_U \to 0 ,$$

where Q and P are direct sums of sheaves $R_V (V \subset U)$. From (7) and (11) we get the commutative diagram

$$\text{(12)} \qquad \begin{array}{ccc} 0 & & 0 \\ \downarrow & & \downarrow \\ \mathrm{Hom}(\Gamma_c(A_U \otimes K), N) & \xrightarrow{\mu_A} & \mathrm{Hom}(A_U, E(R_U)) \\ \downarrow & & \downarrow \beta \\ \mathrm{Hom}(\Gamma_c(P \otimes K), N) & \xrightarrow{\mu_P} & \mathrm{Hom}(P, E(R_U)) \\ \downarrow & & \downarrow \gamma \\ \mathrm{Hom}(\Gamma_c(Q \otimes K), N) & \xrightarrow{\mu_Q} & \mathrm{Hom}(Q, E(R_U)) \end{array}$$

where the columns are exact, as follows from the left exactness of Hom(,N) and from 6.6 (for the left hand column). We have already seen that μ_P and μ_Q are isomorphisms. Simple diagram chasing then shows that μ_A is an isomorphism. This proves the first part of 7.5 . Assume now N to be injective. Then 6.6 implies that $A \mapsto E(A)$ is an exact functor. Together with (1) , this shows that $E(R_X)$ is injective.

7.6 COROLLARY . *The dualizing complex of sheaves* $\mathcal{D}_X^{\bullet}$ *is injective.*

Indeed, in each degree it is a finite direct sum of sheaves of the form $E(R_X)$,(for various K's and injective N's).

7.7 *The dual of a* DGS . We now carry the foregoing over to complexes of sheaves. Given $A^{\bullet} \in \mathrm{DGS}(X)$ we define the dual complex $\mathbf{D}_X^{\bullet} A^{\bullet}$, or simply $\mathbb{D}^{\bullet}A^{\bullet}$, by means of the presheaf

(1) $$U \longmapsto \mathrm{Hom}^{\bullet}(\Gamma_c((A^{\bullet} \otimes K^{\bullet})_U), I^{\bullet}) \ ,$$

which is again a DGS. We have then, as a generalisation of 7.1(3) .

(2) $$\mathbb{H}^{\bullet}(U;\mathbb{D}_X^{\bullet}A^{\bullet}) = \mathrm{Ext}^{\bullet}(\Gamma_c((A^{\bullet} \otimes K^{\bullet})_U), R)$$

(U open in X). The right hand side of (2) is the abutment of a spectral sequence in which

(3) $$E_2^{p,q} = \mathrm{Ext}^p(\mathbb{H}_c^{-q}(U;A^{\bullet}), R) \qquad (p,q \in \mathbb{Z})$$

This also yields analogues of 7.1(5),(6) . If X satisfies the first axiom of countability and $A^{\bullet}$ is cc , then 7.1(7),(8),(9) also extend; we have

(4) $$(\mathbb{D}_X^{\bullet}A^{\bullet})_x = \mathrm{RHom}^{\bullet}(f_x^!A^{\bullet}, R)$$

(5) $$H^{\bullet}(\mathbb{D}_X^{\bullet}A^{\bullet})_x = \mathrm{Ext}^{\bullet}(f_x^!A^{\bullet}, R) \ ,$$

where the right hand side can be computed by using $K^{\bullet} \otimes A^{\bullet}$ or any bounded injective resolution of $A^{\bullet}$. The right hand side is the abutment of a spectral sequence in which

(6) $$E_2^{p,q} = \mathrm{Ext}^p(H^{-q}(f_x^!A^{\bullet}), R) \qquad (p,q \in \mathbb{Z})$$

Note that $\mathbb{D}_X^{\bullet}A^{\bullet}$ can be computed in D(X) by the formula

$$U \longmapsto \mathrm{Hom}^{\bullet}(\Gamma_c(S_U^{\bullet}), I^{\bullet})$$

for any c-soft resolution $A^{\bullet} \to S^{\bullet}$ of $A^{\bullet}$. The advantage of (1) is that it yields a specific element of DGS(X) .

The above definition of the dual complex $\mathbb{D}_X^{\bullet}A^{\bullet}$ follows [2] . On the other hand, the *Verdier dual* of $A^{\bullet}$ in D^bX is $R\mathcal{H}om^{\bullet}(A^{\bullet}, \mathcal{D}_X^{\bullet})$. The equality

(7) $$\mathbb{D}_X^{\bullet}A^{\bullet} = R\mathcal{H}om^{\bullet}(A^{\bullet}, \mathcal{D}_X^{\bullet}) \ , \text{ in } D^b(X) \ ,$$

is a special case of Verdier duality (7.16). The equality (8) is a consequence of the following statement which is more precise in the sense that it asserts the existence of an isomorphism already in $DGS^b(X)$.

7.8 THEOREM . (i) *The functor* $A^\bullet \longmapsto \mathbb{D}_X^\bullet A^\bullet$ *in* DGS(X) *is exact and preserves quasi-isomorphisms.*

(ii) *There is a natural isomorphism in* $DGS^b(X)$:

$$(1) \qquad \mathbb{D}_X^\bullet A^\bullet \tilde{\to} \mathcal{H}om^\bullet(A^\bullet, \mathcal{D}_X^\bullet) , \qquad (A^\bullet \in DGS^b(X)) .$$

Proof : The first assertion of (i) follows from 7.5 . Let $\alpha : A^\bullet \to B^\bullet$ be a q.i.. By 1.4 , it yields an isomorphism $\mathbb{H}_c^\bullet(U;A^\bullet) \to \mathbb{H}_c^\bullet(U,B^\bullet)$, (U open in X), hence also an isomorphism of the E_2-terms, given by 7.7(3) of the spectral sequences abuting to the right-hand sides of 7.7(2) for $A^\bullet$ and $B^\bullet$ respectively. Thus α yields an isomorphism of $\mathbb{H}^\bullet(U;\mathbb{D}_X^\bullet B^\bullet)$ onto $\mathbb{H}^\bullet(U;\mathbb{D}_X^\bullet A^\bullet)$. Going over to the limit over a fundamental set of neighborhoods of a point $x \in X$, we get then an isomorphism of $\mathcal{H}^\bullet(\mathbb{D}_X^\bullet B^\bullet)_x$ onto $\mathcal{H}^\bullet(\mathbb{D}_X^\bullet A^\bullet)_x$. This proves the second part of (i) . The map μ of 7.5(1) , applied to A^i, K^j, I^k, $(i,j,k \in \mathbb{Z})$ yields a map

$$(2) \qquad \mathbb{D}_X^\bullet A^\bullet \longrightarrow \mathcal{H}om^\bullet(A^\bullet, \mathcal{D}_X^\bullet) ,$$

which is bijective by 7.5 and obviously commutes with the differentials. This gives (ii).

7.9 *A complement to* II,7.4 *in* [5] . Before specializing the foregoing we make some remaks about homomorphisms of sheaves.

Let $L, M \in Sh(X)$. Given $x \in X$, there is a natural homomorphism

$$(1) \qquad e_x : \mathcal{H}om(L,M)_x \longrightarrow \mathrm{Hom}(L_x, M_x) ,$$

and more generally

$$(2) \qquad e_x^\bullet : \mathcal{E}xt^\bullet(L,M)_x \longrightarrow \mathrm{Ext}^\bullet(L_x, M_x) .$$

As is well known e_x is neither surjective nor injective in general. We have however [5:II,7.4] :

If L is locally constant with finitely generated stalks, then we can compute $\mathbb{R}Hom^{\bullet}(L,M)$ *and* $Ext^{\bullet}(L,M)$ *by using locally a left resolution of* L *by finitely generated free sheaves (instead of a right injective resolution of* M *, as is always possible). Moreover* $e_x^{\bullet}$ *is an isomorphism.*

The key remark here is that the functor $M \longmapsto Hom(L,M)$ is exact if L is locally free of finite rank (this is a consequence of the equality $Hom(R_X,M) = M)$, and the general case follows then by standard homological algebra. To handle clc complexes, we shall need the following variant.

LEMMA . *Let* X *be locally contractible and* $L,M \in Sh(X)$ *be locally constant. Then we can compute* $\mathbb{R}Hom^{\bullet}(L,M)$ *and* $Ext^{\bullet}(L,M)$ *by using locally a free left resolution of* L *. Moreover* $Ext^i(L,M)$ *is locally constant and* $e_x^{\bullet}$ *is an isomorphism.*

We check this only in the case where $L = \bigoplus_{i \in I} L_i$ with $L_i = R_X$ and leave it to the reader to conclude as in [5:II,7.4].
We have

$$Hom(L,M) = M^I . \tag{3}$$

Since X is locally connected, M^I is locally constant and

$$(M^I)_x = (M_x)^I = \mathrm{Hom}(L_x,M_x) , \quad (x \in X) .$$

It remains only to check that M is acyclic for $Hom(L,\ )$.(Note here that the functor $A \longrightarrow A^I$ is in general not exact if I is not finite, since an arbitrary product of exact sequences in Sh(X) need not be exact) . Let $J^{\bullet}$ be an injective resolution of M . We must show that

$$Hom(L,M) \longrightarrow Hom^{\bullet}(L,J^{\bullet})$$

is still a resolution. It is sufficient to show that if U is a contractible open subset of X, then the sequence of R-modules

$$0 \to M(U)^I \to J^0(U)^I \to J^1(U)^I \to \dots \tag{4}$$

is exact. But this can be checked componentwise, and the exactness of

(5) $$0 \to \mathcal{M}(U) \to \mathcal{J}^0(U) \to \mathcal{J}^1(U) \to \dots$$

is equivalent to the vanishing of $H^i(U;\mathcal{M})$ for $i > 0$, which follows from 1.11(b).

7.10 *A special case.* Let X be a manifold, $n = \dim X$ and $\mathcal{O}$ the orientation sheaf of X. We have seen that $\mathcal{D}_X^\bullet[-n]$ is an injective resolution of $\mathcal{O}$. (7.3). Therefore, if $A^\bullet \in DGS(X)$, then

(1) $$\mathbf{D}^\bullet A^\bullet[-n] = R\mathcal{H}om^\bullet(A^\bullet,\mathcal{O})$$

and therefore

(2) $$\mathcal{H}^{i-n}(\mathbf{D}^\bullet A^\bullet)_x = \mathcal{E}xt^i(A^\bullet,\mathcal{O})_x \quad (i \in \mathbb{Z}\ ;\ x \in X)$$

Assume now that $A^\bullet = \mathcal{E}$ is a locally constant sheaf. Then (2) and 7.9 imply :

(3) $$\mathcal{H}^{i-n}(\mathbf{D}^\bullet\mathcal{E})_x = \mathrm{Ext}^i(\mathcal{E}_x,R)\ , \quad (x \in X;\ \ i \in \mathbb{Z}).$$

It follows in particular that $\mathbf{D}^\bullet\mathcal{E}$ is clc, with finitely generated stalk cohomology if $\mathcal{E}_x$ moreover is finitely generated, and that x has a neighborhood on which $\mathbf{D}^\bullet\mathcal{E}[-n]$ is q.i. to the complex of constant sheaves with stalks $\mathrm{Hom}^\bullet(\mathcal{E}_x,I^\bullet)$, where $I^\bullet$ is an injective resolution of R. From (3) we also get

(4) $$\mathbf{D}^\bullet\mathcal{E}[-n] = \mathcal{E}^* \otimes \mathcal{O}\ , \text{ if } \mathcal{E}_x \text{ is a free R-module } (x \in X)\ ,$$

where $\mathcal{E}^*$ is the locally constant sheaf $\mathcal{H}om(\mathcal{E},R_X)$ with stalks $\mathcal{E}^*_x = \mathrm{Hom}(\mathcal{E}_x,R)$.

C. *The functors $f_!$ and $f^!$. Verdier duality.*

We now discuss Verdier duality with respect to a continuous map $f : X \to Y$. We refer to VI for a more detailed exposition of the basic facts on $f_!$ and $f^!$.

7.11 *The functor direct image with proper support* $f_!$.
Given a sheaf A on X , consider the presheaf on Y defined by

(1) $$V \mapsto \Gamma_{\Phi_V}(f^{-1}V;A) , \qquad (V \text{ open in } Y)$$

where Φ_V is the family of closed subsets C of $f^{-1}V$ such that $f : C \to V$ is a proper map. It is not difficult to see that if $\{V_i\}$ is an open cover of V and C is closed in $f^{-1}V$, then $f : C \to V$ is proper if and only if the maps $f : f^{-1}V_i \cap C \to V_i$ are proper. From this it follows that the presheaf (1) is a sheaf, contained in f_*A , to be denoted $f_!A$ (see VI,2.2 or [8:p.119]) .
We have then

(2) $$f_!A(V) = \Gamma_{\Phi_V}(f^{-1}V;A) , \qquad (V \text{ open in } Y) .$$

In particular, if f is a proper map, then $f_!$ is the direct image functor f_*(1.6); if Y is a point, then $f_!A = \Gamma_c(X;A)$.

7.12 LEMMA . *Let* K *be a c-soft sheaf on* X . *Then for every* $y \in Y$ *we have a canonical isomorphism*

(1) $$\alpha : f_!(K)_y = \Gamma_c(f^{-1}y;K|f^{-1}y)$$

Proof : If V is an open neighborhood of y , then the intersection of any $C \in \Phi_V$ with $f^{-1}y$ is compact, hence the restriction gives a map:

(2) $$\alpha : f_!(K)(V) \to \Gamma_c(f^{-1}y;K) .$$

Since K is c-soft, the map α is surjective. On the other hand, if an element s of the left hand side restricts to zero, then its support C is a closed subset of $f^{-1}V$ not meeting $f^{-1}y$, therefore f(C) is a closed subset of V not containing y . Then $s_y = 0$, which shows that α is injective.

Remark. 7.12 is true for any sheaf (see VI,2.6) but we shall not need this fact.

7.13 Fix $B \in Sh(Y)$ and a c-soft flat sheaf K on X. For $A \in Sh(X)$, set

(1) $$\mathcal{E}(A) = \mathcal{H}om(f_!(A \otimes \mathcal{K}), \mathcal{B}) .$$

We have then the

LEMMA . *Assume* $\mathcal{B}$ *to be injective . Then the functors* $A \mapsto f_!(A \otimes \mathcal{K})$ *and* $A \mapsto \mathcal{E}(A)$ *are exact and* $\mathcal{E}(A)$ *is flabby .*

Proof . The sheaf $A \otimes \mathcal{K}$ is c-soft (6.5), hence so is its restriction to $f^{-1}(y)$. In view of 6.6 , it follows that $A \mapsto \Gamma_c(A \otimes \mathcal{K}|f^{-1}y)$ is exact. By 7.12 , applied to $A \otimes \mathcal{K}$, this implies that $A \mapsto f_!(A \otimes \mathcal{K})$ is exact. Since $\mathcal{B}$ is injective, the first assertion follows. The second one is standard (cf.[5], lemma 7.3.2, p.264) .

7.14 *The functor* $f^!$. Given a sheaf $\mathcal{B}$ on Y , the rule

(1) $$U \mapsto \mathrm{Hom}(f_!\mathcal{K}_U, \mathcal{B}) , \quad (U \text{ open in } X)$$

defines a sheaf on X , to be denoted by $f^!_{\mathcal{K}}\mathcal{B}$. We claim there is a canonical isomorphism

(2) $$\mathcal{H}om(f_!(A \otimes \mathcal{K}), \mathcal{B}) = \mathcal{E}(A) = f_*\mathcal{H}om(A, f^!_{\mathcal{K}}\mathcal{B}) .$$

If Y is a point, then $f_!(A \otimes \mathcal{K}) = \Gamma_c(A \otimes \mathcal{K})$, therefore (2) , used simultaneously for all the sheaves A_U (U open in X) , gives 7.5(1) back.

To prove (2) in general, we consider the two functors from $Sh(X) \times Sh(Y)$ to $Sh(Y)$ given by

$$A, \mathcal{B} \mapsto E_{\mathcal{B}}(A) = \mathcal{H}om(f_!(A \otimes \mathcal{K}), \mathcal{B})$$

$$(A \in Sh(X),\ \mathcal{B} \in Sh(Y))$$

$$A, \mathcal{B} \mapsto F_{\mathcal{B}}(A) = f_*\mathcal{H}om(A, f^!_{\mathcal{K}}\mathcal{B})$$

We have to prove that they are naturally isomorphic. Fix A and consider them as functors in $\mathcal{B}$. They are left exact and commute with arbitrary direct products. If supp $\mathcal{B}$ is a point, then it is easily seen that we are reduced to 7.5 . We have then also an isomorphism $E_{\mathcal{B}}(A) \tilde{\to} F_{\mathcal{B}}(A)$ when $\mathcal{B}$ is a product of skyscraper sheaves. In general, consider the beginning $0 \to \mathcal{B} \to \mathcal{P} \to \mathcal{Q}$ of the canonical flabby resolution of $\mathcal{B}$. We get a commutative diagram

$$\begin{array}{ccccccc} 0 & \longrightarrow & E_{\mathcal{B}}(A) & \longrightarrow & E_{\mathcal{P}}(A) & \longrightarrow & E_{\mathcal{Q}}(A) \\ & & & & \downarrow & & \downarrow \\ 0 & \longrightarrow & F_{\mathcal{B}}(A) & \longrightarrow & F_{\mathcal{P}}(A) & \longrightarrow & F_{\mathcal{Q}}(A) \end{array}$$

with exact rows in which the two vertical arrows are isomorphisms since $\mathcal{P}$ and $\mathcal{Q}$ are products of skyscraper sheaves. This yields a canonical isomorphism $E_{\mathcal{B}}(A) \to F_{\mathcal{B}}(A)$, whence (2).

7.15 Now assume $\mathcal{B}$ to be injective. Taking global sections in 7.14(2), we get the isomorphism

(1) $$\mathrm{Hom}(f_!(A \otimes K), \mathcal{B}) = \mathrm{Hom}(A, f^!_K\mathcal{B}) .$$

Since $f_!(A \otimes K)$ is exact in A (7.13), we see that the right hand side is also exact in A, hence $f^!_K\mathcal{B}$ is injective, i.e. $f^!_K$ transforms injective sheaves into injective ones.

7.16 Let now $\mathcal{J}^\bullet \in \mathrm{DGS}(X)$. We let $f^!_K.\mathcal{J}^\bullet$ be the simple complex associated to the double complex $\{f^!_{K^i}\mathcal{J}^j\}$ $(i,j \in \mathbb{Z})$. For $A^\bullet \in \mathrm{DGS}(X)$, 7.14 yields an isomorphism in $\mathrm{DGS}(Y)$:

(2) $$\mathcal{H}om^\bullet(f_!(A^\bullet \otimes K^\bullet), \mathcal{J}^\bullet) = f_*\mathcal{H}om^\bullet(A^\bullet, f^!_K.\mathcal{J}^\bullet) .$$

Let now $\mathcal{B}^\bullet \in \mathrm{DGS}^b(Y)$; define $f^!\mathcal{B}^\bullet$ in $D^b(X)$ by $f^!_K.\mathcal{J}^\bullet$ where $\mathcal{J}^\bullet$ is an injective resolution of $\mathcal{B}^\bullet$, i.e. $f^!\mathcal{B}^\bullet = Rf^!_K.\mathcal{B}^\bullet$. It consists of injective sheaves. Moreover, $\mathcal{H}om^\bullet(A^\bullet, f^!_K.\mathcal{J}^\bullet)$ is flabby.
Therefore $f_*\mathcal{H}om^\bullet(A^\bullet, f^!\mathcal{B})$ represents $Rf_*R\mathcal{H}om^\bullet(A^\bullet, f^!\mathcal{B})$. On the other hand $A^\bullet \otimes K^\bullet$ is a c-soft resolution of $A^\bullet$, hence $Rf_!A^\bullet = f_!(A^\bullet \otimes K^\bullet)$ and the left hand side of (2) represents $R\mathcal{H}om^\bullet(Rf_!A^\bullet, \mathcal{B}^\bullet)$. This proves therefore:

7.17 THEOREM (Verdier duality). *In* $D^b(Y)$, *we have the canonical isomorphism*

$$\text{(1)} \qquad R\mathcal{H}om^{\bullet}(Rf_! A^{\bullet}, \mathcal{B}^{\bullet}) = Rf_* R\mathcal{H}om^{\bullet}(A^{\bullet}, f^! \mathcal{B}^{\bullet})$$

This theorem was announced by J.L. Verdier in [11]. A proof is given in [10:Exp.4]. The notes [8] stop short of the statement 7.16 but give all the essentials of the proof (see IV.6 and V.4 there). We have to a large extent followed the exposition of [8].

7.18 Assume now that Y is a point and $\mathcal{B}^{\bullet} = R_{pt}$. Then $\mathcal{J}^{\bullet}$ is just an injective resolution of R . Take first $A^{\bullet} = R_X$. Then the left hand side is $\mathcal{D}_X^{\bullet}$, while the right hand side is $f^! R_{pt}$. Hence

$$\text{(2)} \qquad \mathcal{D}_X^{\bullet} = f^! R_{pt} \text{ , if } f : X \to pt .$$

For a general $A^{\bullet}$, 7.17 follows in the present case from 7.5 .

7.19 If $S^{\bullet}, T^{\bullet} \in D^b(Z)$, we have by 5.17(3)

$$\text{Mor}_{D^b(Z)}(S^{\bullet}, T^{\bullet}) = \mathbf{H}^0(Z; R\mathcal{H}om^{\bullet}(S^{\bullet}, T^{\bullet})) .$$

It follows then from 7.17 that

$$\text{(1)} \qquad \text{Mor}_{D^b(X)}(A^{\bullet}, f^! \mathcal{B}^{\bullet}) = \text{Mor}_{D^b(Y)}(Rf_! A^{\bullet}, \mathcal{B}^{\bullet}) .$$

Suppose now that $f : X \to Y$ is the inclusion of a closed subset. In 1.8 we have defined a functor $\gamma_X : Sh(Y) \to Sh(X)$. It is right adjoint to $f_!$, and it follows easily that $R\gamma_X$ is right adjoint to $Rf_!$. Thus the functor $R\gamma_X$ and $f^!$ are both right adjoint to $Rf_!$, and they are therefore canonically isomorphic. As a consequence the definition of $f^!$ given in 1.8 is compatible with the one given here.

This can also be checked directly as follows. For $\mathcal{B}^{\bullet} \in D^b(Y)$ and $U \subset X$ open we have

$$f^! \mathcal{B}^{\bullet}(U) = \text{Hom}^{\bullet}(f_! \mathcal{K}_U^{\bullet}, \mathcal{B}^{\bullet}) = \text{Hom}^{\bullet}(\mathcal{K}_U^{\bullet}, \gamma_X \mathcal{B}^{\bullet})$$

Using the natural map $R_U \to \mathcal{K}_U^{\bullet}$, we get a morphism

$$\phi_U : \text{Hom}^{\bullet}(\mathcal{K}_U^{\bullet}, \gamma_X \mathcal{B}^{\bullet}) \to \text{Hom}^{\bullet}(R_U, \gamma_X \mathcal{B}^{\bullet}) = \gamma_X \mathcal{B}^{\bullet}(U) ,$$

whence a morphism $\phi : f^{!}\mathcal{B}^{\cdot} \to \gamma_X\mathcal{B}^{\cdot}$. To check that ϕ is a q.i., it suffices to notice that each ϕ_U is a q.i. since $R_U \to K^{\cdot}_U$ is one and $\gamma_X\mathcal{B}^{\cdot}$ is injective.

A similar argument shows that $f^{!} = f^*$, if f is the inclusion of an open subset.

§ 8 CONSTRUCTIBILITY OF R*Hom* AND BIDUALITY ON A PSEUDOMANIFOLD

A. *Constructibility*

8.1 LEMMA . *Let* Y *be a locally compact space, every point of which has a fundamental system of contractible neighborhoods and* $S^\bullet$ *a clc complex of sheaves on* Y . *Then every point* $y \in Y$ *has a neighborhood on which* $S^\bullet$ *is q.i. to a complex of constant sheaves.*

Proof. Let U be an open contractible subspace of Y . The sheaf $\mathcal{H}^\bullet S^\bullet$ is constant on U . The E_2 term of the hypercohomology spectral sequence on U with respect to $S^\bullet$ is given by :

$$(1) \qquad E_2^{p,q} = H^p(U;\mathcal{H}^q S^\bullet) = \begin{cases} 0 & \text{if } p \neq 0 , \\ \mathcal{H}^q S^\bullet_y & \text{if } p = 0 \text{ and } y \in U . \end{cases}$$

(we have used 1.11(b)). We have then

$$(2) \qquad \mathbb{H}^q(U;S^\bullet) = E_2^{0,q} = \mathcal{H}^q S^\bullet_y \ , \quad (q \in \mathbb{Z}; y \in U) .$$

Let $S^\bullet \to J^\bullet$ be an injective resolution of $S^\bullet$ and let $T^\bullet$ be the complex of constant sheaves on U such that

$$(3) \qquad T^p_y = \Gamma(U;J^p) \ , \quad (p \in \mathbb{Z})$$

The the restriction supplies a morphism $T^\bullet \to J^\bullet$ which is obviously a q.i..

8.2 PROPOSITION . *Let* Y *be a locally compact space,* Z *a closed subspace and* $j : Z \to Y$ *the inclusion map. Then, for any* $S^\bullet \in \mathrm{DGS}(Y)$, *we have in* D(Z) *the isomorphism*

$$(1) \qquad \mathbb{D}^\bullet_Z j^* S^\bullet = j^! \mathbb{D}^\bullet_Y S^\bullet .$$

Proof : Let $K^\bullet$ be a c-soft flat resolution of R_Y . Then $j^*K^\bullet$ is a c-soft flat resolution of R_Z . If we use these resolutions in 7.7(1) and set $j^!\mathbb{D}_Y^\bullet S^\bullet = \gamma_Z(\mathbb{D}_Y^\bullet S^\bullet)$ (this is allowed since $\mathbb{D}_Y^\bullet S^\bullet$ is flabby), then the isomorphism (1) actually holds in DGS(Z) . We prove this stronger statement.

For each open subset $V \subset Y$, we have an exact sequence

$$(2) \qquad 0 \to \Gamma_c(V-Z;S^\bullet \otimes K^\bullet) \to \Gamma_c(V;S^\bullet \otimes K^\bullet) \to \Gamma_c(Z\cap V;j^*S^\bullet \otimes j^*K^\bullet) \to 0$$

Applying $\mathrm{Hom}^\bullet(\ \ ,I^\bullet)$ to (2), we get an exact sequence

$$(3) \qquad 0 \leftarrow \Gamma(V-Z;\mathbb{D}_Y^\bullet S^\bullet) \leftarrow \Gamma(V;\mathbb{D}_Y^\bullet S^\bullet) \leftarrow \Gamma(Z\cap V;\mathbb{D}_Z^\bullet j^*S^\bullet) \leftarrow 0$$

hence an exact sequence

$$(4) \qquad 0 \leftarrow i_*i^*\mathbb{D}_Y^\bullet S^\bullet \leftarrow \mathbb{D}_Y^\bullet S^\bullet \leftarrow j_!\mathbb{D}_Z^\bullet j^*S^\bullet \leftarrow 0$$

But since $\mathbb{D}_Y^\bullet S^\bullet$ is flabby we know that the kernel of $\mathbb{D}_Y^\bullet S^\bullet \to i_*i^*\mathbb{D}_Y^\bullet S^\bullet$ is $j_!\gamma_Z\mathbb{D}_Y^\bullet S^\bullet = j_!j^!\mathbb{D}_Y^\bullet S^\bullet$. Thus $\mathbb{D}_Z^\bullet j^*S^\bullet = j^!\mathbb{D}_Y^\bullet S^\bullet$.

8.3 PROPOSITION . *Let* X *be a pseudomanifold and* $\mathfrak{X}$ *an unrestricted (see* 2.1*) stratification of* X . *Then* $\mathcal{D}_X^\bullet$ *is* $\mathfrak{X}$ - *cc* .

Proof : We use the notation 2.1 for $\mathfrak{X}$ and prove by induction on $k \geq 1$ that $(\mathcal{D}_X^\bullet)_k = \mathcal{D}_{U_k}^\bullet$ is $\mathfrak{X}$ - cc . On U_1 , the complex $\mathcal{D}_X^\bullet$ is q.i. to the orientation sheaf, up to a shift (7.10) whence our assertion in this case. Assume it is proved for some $k \geq 1$. Consider the exact triangle

$$(1) \qquad \begin{array}{ccc} j_{k!}j_k^!\mathcal{D}_X^\bullet & \longrightarrow & \mathcal{D}_{U_{k+1}}^\bullet \\ {\scriptstyle [1]}\nwarrow & & \swarrow \\ & Ri_{k*}\mathcal{D}_{U_k}^\bullet & \end{array}$$

By induction and 3.11 , $Ri_{k*}\mathcal{D}^{\bullet}_{U_k}$ is $\mathfrak{X}$ - cc on U_{k+1} . On the other hand, 8.2 gives

$$j_k^{!}\mathcal{D}^{\bullet}_X = j_k^{!}\mathbf{D}^{\bullet}_X(R_X) = \mathbf{D}^{\bullet}_{S_{k+1}} j_k^{*}R_X = \mathbf{D}^{\bullet}_{S_{k+1}} R_{S_{k+1}} = \mathcal{D}^{\bullet}_{S_{k+1}} ,$$

which is clc with finitely generated stalked cohomology, since S_{k+1} is a manifold. It follows that $\mathcal{D}^{\bullet}_{U_{k+1}}$ is $\mathfrak{X}$ - cc (5.13) .

8.4 We now discuss the contructibility of $R\mathcal{H}om$. We let D(R) denote the derived category of R-modules.

If Y is a space and $\mathcal{S}^{\bullet},\mathcal{T}^{\bullet} \in DGS(Y)$, then the map $e_x (x \in Y)$ of 7.9 extends naturally to a morphism

(1) $$Re^{\bullet}_x : R\mathcal{H}om^{\bullet}(\mathcal{S}^{\bullet},\mathcal{T}^{\bullet})_x \to RHom^{\bullet}_{D(R)}(\mathcal{S}^{\bullet}_x,\mathcal{T}^{\bullet}_x) , (x \in Y) .$$

We are interested in giving conditions under which $Re^{\bullet}_x$ is an isomorphism. We recall [5:I,5.4.1] that the cohomology of the right hand side is the abutment of a spectral sequence in which

(2) $$E_2^{p,q} = \oplus_i Ext^p(H^{-i}(\mathcal{S}^{\bullet}_x) , H^{q+i}(\mathcal{T}^{\bullet}_x)) , (p,q \in \mathbb{Z}) .$$

8.5 PROPOSITION . *Let* Y *be a locally contractible space and* $\mathcal{L}^{\bullet},\mathcal{M}^{\bullet} \in DGS(Y)$ *be clc . Then*

(1) $$R\mathcal{H}om^{\bullet}(\mathcal{L}^{\bullet},\mathcal{M}^{\bullet})_x = RHom^{\bullet}_{D(R)}(\mathcal{L}^{\bullet}_x,\mathcal{M}^{\bullet}_x) \quad (x \in Y)$$

and $RHom^{\bullet}_{D(Y)}(\mathcal{L}^{\bullet},\mathcal{M}^{\bullet})$ *is clc . The latter has finitely generated stalk cohomology if* $\mathcal{L}^{\bullet}$ *and* $\mathcal{M}^{\bullet}$ *have that property .*

Proof . Since our assertions are local, we may, in view of 8.1 , assume $\mathcal{L}^{\bullet}$ and $\mathcal{M}^{\bullet}$ to consist of constant sheaves. Then $\mathcal{L}^{\bullet}$ has a left resolution $\mathcal{C}^{\bullet} \to \mathcal{L}^{\bullet}$ by constant sheaves with R-free stalks, and $R\mathcal{H}om^{\bullet}(\mathcal{L}^{\bullet},\mathcal{M}^{\bullet})$ is q.i. to $R\mathcal{H}om^{\bullet}(\mathcal{C}^{\bullet},\mathcal{M}^{\bullet})$. The arguments of 7.9 show that we can take

(2) $$R\mathcal{H}om^{\bullet}(L^{\bullet},M^{\bullet}) = \mathcal{H}om^{\bullet}(C^{\bullet},M^{\bullet})$$

and that

(3) $$\mathcal{H}om^{\bullet}(C^{\bullet},M^{\bullet})_x = \mathrm{Hom}^{\bullet}(C^{\bullet}_x,M^{\bullet}_x) .$$

In particular this gives (1). Using the fact that Y is locally connected, we find also that $\mathcal{H}om^{\bullet}(C^{\bullet},M^{\bullet})$ consists of constant sheaves and that it is clc . Thus $R\mathcal{H}om^{\bullet}(L^{\bullet},M^{\bullet})$ is clc . The last assertion of the proposition follows from (1) .

8.6 THEOREM . *Let* X *be a pseudomanifold and* $\mathfrak{X}$ *an unrestricted stratification of* X . *Let* $A^{\bullet},B^{\bullet} \in \mathrm{DGS}(X)$ *be bounded below and* $\mathfrak{X}$ *- clc (resp.* $\mathfrak{X}$ *- cc). Then* $R\mathcal{H}om^{\bullet}(A^{\bullet},B^{\bullet})$ *is* $\mathfrak{X}$ *- clc (resp.* $\mathfrak{X}$ *- cc).*

Proof : In the notation of 2.1 , we want to prove by induction on $k \geq 1$ that $R\mathcal{H}om^{\bullet}(A^{\bullet},B^{\bullet})_k$ is $\mathfrak{X}$ - clc (resp. $\mathfrak{X}$ - cc) if $A^{\bullet}$ and $B^{\bullet}$ are so. For $k = 1$, this follows from 8.5 . Assume it is proved for some $k \geq 1$. If we apply $R\mathcal{H}om^{\bullet}(\ ,B^{\bullet})$ to the distinguished triangle

(1) 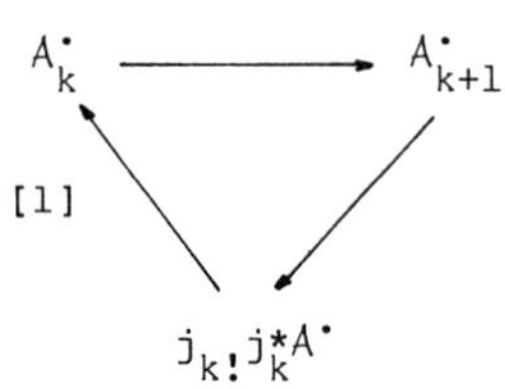

we get a triangle

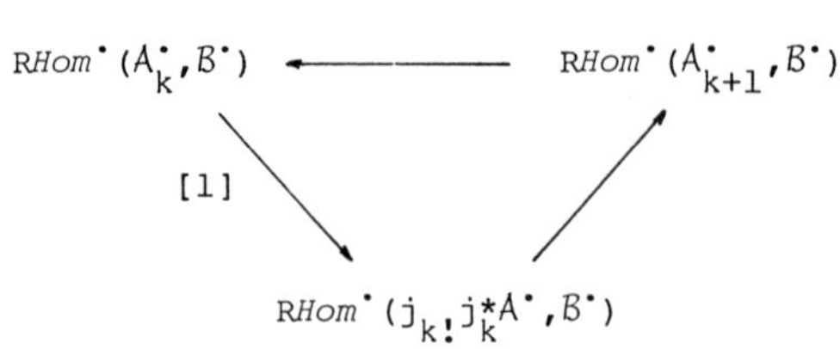

which is also distinguished (5.10).By 5.13, it suffices to show that two vertices are $\mathfrak{X}$ - clc (resp. $\mathfrak{X}$ - cc).
It follows immediately from the definitions that

(3) $$R\mathcal{H}om^\bullet(i_{k!}A_k^\bullet, \mathcal{B}^\bullet) = Ri_{k*}R\mathcal{H}om^\bullet(A_k^\bullet, \mathcal{B}_k^\bullet) .$$

By induction assumption $R\mathcal{H}om^\bullet(A_k^\bullet, \mathcal{B}_k^\bullet)$ is $\mathfrak{X}$ - clc (resp. $\mathfrak{X}$ - cc) on U_k . Then the right hand side of (3) is $\mathfrak{X}$ - clc (resp. $\mathfrak{X}$ - cc) on U_{k+1} by 3.9 (resp. 3.11).

Verdier duality (7.17), applied to j_k , gives

(4) $$R\mathcal{H}om^\bullet(j_{k!}j_k^*A^\bullet, \mathcal{B}^\bullet) = Rj_{k*}R\mathcal{H}om^\bullet(j_k^*A^\bullet, j_k^!\mathcal{B}^\bullet) .$$

On S_{k+1} , the complex $j_k^!\mathcal{B}^\bullet$ is clc (resp. clc with finitely generated stalk cohomology) if $\mathcal{B}^\bullet$ is $\mathfrak{X}$ - clc (resp. $\mathfrak{X}$ - cc) by 3.10 , therefore $R\mathcal{H}om^\bullet(j_k^*A^\bullet, j_k^!\mathcal{B}^\bullet)$ is clc (resp. clc with finitely generated stalk cohomology) by 8.5 . The same is then true for $Rj_{k*}R\mathcal{H}om^\bullet(j_k^*A^\bullet, j_k^!\mathcal{B}^\bullet)$, which is just extension by zero.

8.7 COROLLARY . *Let $S^\bullet \in DGS(X)$ be $\mathfrak{X}$ - clc (resp. $\mathfrak{X}$ - cc). Then $\mathbb{D}_X^\bullet S^\bullet$ is $\mathfrak{X}$ - clc (resp. $\mathfrak{X}$ - cc).*

We have $\mathbb{D}_X^\bullet S^\bullet = R\mathcal{H}om^\bullet(S^\bullet, \mathcal{D}_X^\bullet)$ by 7.8 and $\mathcal{D}_X^\bullet$ is $\mathfrak{X}$ - cc by 8.4 . We may therefore apply 8.6 .

B. Biduality

8.8 PROPOSITION . *Let Z be a closed subspace of the locally compact space Y . Let $j : Z \to Y$ and $i : U = Y - Z \to Y$ be the inclusion maps. We have then in D(Y) :*

(1) $$\mathbb{D}_Y^\bullet j_! A^\bullet = j_! \mathbb{D}_Z^\bullet A^\bullet$$

$(A^\bullet \in DGS(Z);\ \mathcal{B}^\bullet \in DGS(U))$

(2) $$\mathbb{D}_Y^\bullet i_! \mathcal{B}^\bullet = Ri_* \mathbb{D}_U^\bullet \mathcal{B}^\bullet$$

Assume moreover that Y satisfies the first axiom of countability and that $Ri_*\mathcal{B}^\bullet$ is cc ($\mathcal{B}^\bullet \in DGS(U)$). Then

(3) $$\mathbb{D}_Y^\bullet Ri_* \mathcal{B}^\bullet = i_! \mathbb{D}_U^\bullet \mathcal{B}^\bullet .$$

Proof : Let $\mathcal{K}^\bullet$ be the c-soft flat resolution of R_Y used to define $\mathcal{D}_Y^\bullet$. Then $j^*\mathcal{K}^\bullet$ and $i^*\mathcal{K}^\bullet$ are c-soft flat resolutions of R_Z and R_U respectively . We use them to compute $\mathbf{D}_Z^\bullet$ and $\mathbf{D}_U^\bullet$. Since $\mathbf{D}_U^\bullet\mathcal{B}^\bullet$ is flabby, we may set $Ri_*\mathbf{D}_U^\bullet\mathcal{B}^\bullet = i_*\mathbf{D}_U^\bullet\mathcal{B}^\bullet$. We claim that with these choices (1) and (2) hold already in DGS(Y) . Indeed, these stronger statements follow from the definitions (7.7) and from the fact that, for $V \subset Y$ open, we have

(4) $$\Gamma_c(j_*\mathcal{A}^\bullet \otimes \mathcal{K}_V^\bullet) = \Gamma_c(\mathcal{A}^\bullet \otimes (j^*\mathcal{K}^\bullet)_{V\cap Z})$$

(5) $$\Gamma_c(i_!\mathcal{B}^\bullet \otimes \mathcal{K}_V^\bullet) = \Gamma_c(\mathcal{B}^\bullet \otimes (i^*\mathcal{K}^\bullet)_{U\cap V}) .$$

This proves (1) and (2) . For (3) , it is clear that $\mathbf{D}_Y^\bullet Ri_*\mathcal{B}^\bullet$ and $i_!\mathbf{D}_U^\bullet\mathcal{B}^\bullet$ agree over U . Thus it is enough to check that $(\mathbf{D}_Y^\bullet Ri_*\mathcal{B}^\bullet)_x = 0$ for $x \in Z$. But the constructibility assumption implies (7.7(6))

(6) $$\mathcal{H}^\bullet(\mathbf{D}_Y^\bullet Ri_*\mathcal{B}^\bullet)_x = \mathrm{Ext}^\bullet(f_x^! Ri_*\mathcal{B}^\bullet, R) .$$

It suffices then to show that $\mathcal{H}^\bullet(f_x^! Ri_*\mathcal{B}^\bullet) = 0$. Since $f_x^! = \ell_x^! \circ j^!$, where ℓ_x is the inclusion of x in Z , it is sufficient to prove that $\mathcal{H}^\bullet(j^! Ri_*\mathcal{B}^\bullet) = 0$. This follows from 1.8(7) applied to $\mathcal{S}^\bullet = i_*\mathcal{J}^\bullet$, where $\mathcal{J}^\bullet$ is an injective resolution of $\mathcal{B}^\bullet$ on U .

8.9 *The map* $BD_X : \mathcal{S}^\bullet \longmapsto \mathbf{D}_X^\bullet\mathbf{D}_X^\bullet\mathcal{S}^\bullet$. Let first $\mathcal{S},\mathcal{T} \in Sh(X)$. There is an obvious sheaf homomorphism $e : \mathcal{S} \to \mathcal{H}om(\mathcal{H}om(\mathcal{S},\mathcal{T}),\mathcal{T})$ which generalizes the canonical map of a module into its bidual : given $U \subset X$ open and $s \in \mathcal{S}(U)$, we have to associate to it a map of $\mathcal{H}om(\mathcal{S},\mathcal{T})|U$ to $\mathcal{T}|U$. By definition, this consists of a collection of maps $\mathrm{Hom}(\mathcal{S}|_V,\mathcal{T}|_V) \to \mathcal{T}(V)$ for all V open in U , compatible with restrictions. Now any $c \in \mathrm{Hom}(\mathcal{S}|_V,\mathcal{T}|_V)$ yields in particular a map $c_V : \mathcal{S}(V) \to \mathcal{T}(V)$. We set

(1) $$e(s)(c) = c_V(s) \; .$$

This extends obviously to graded sheaves. But this yields a morphism in DGS(X) only up to signs. Given now $\mathcal{S}^\bullet, \mathcal{T}^\bullet \in$ DGS(X) , we then define

(2) $$e(s)(c) = (-1)^{ij} c_V(s) \quad (s \in \mathcal{S}^i(U),\ c \in \mathrm{Hom}^j(\mathcal{S}^\bullet|_V, \mathcal{T}^\bullet|_V); i,j \in \mathbb{Z}).$$

We leave it to the reader to check that this is indeed a morphism.

From now on X is a pseudomanifold, $\mathfrak{X}$ an unrestricted (see 2.1) stratification of X and $D^b(\mathfrak{X})$ the derived category of bounded complexes which are $\mathfrak{X}$ - cc . We let $BD_X : \mathcal{S}^\bullet \mapsto \mathfrak{D}^\bullet\mathfrak{D}^\bullet\mathcal{S}^\bullet$ be the map defined by (2) for $\mathcal{T}^\bullet = \mathcal{D}^\bullet_X$. We have seen that $\mathcal{S}^\bullet \mapsto \mathfrak{D}^\bullet\mathcal{S}^\bullet$ preserves $D^b(\mathfrak{X})$ (8.6) and we know by 7.8 that it is exact and preserves q.i.. The same is then true of $\mathcal{S}^\bullet \mapsto \mathfrak{D}^\bullet\mathfrak{D}^\bullet\mathcal{S}^\bullet$.

8.10 THEOREM . *If* $\mathcal{S}^\bullet \in D^b(\mathfrak{X})$, *then* $BD_X : \mathcal{S}^\bullet \longrightarrow \mathfrak{D}^\bullet\mathfrak{D}^\bullet\mathcal{S}^\bullet$ *is an isomorphism in* $D^b(\mathfrak{X})$.

a) We first assume X to be a manifold with the trivial stratification. Then $\mathcal{S}^\bullet$ is clc with finitely generated stalk cohomology . Our assertion being local we may assume that X is a ball of dimension n . Then $\mathcal{H}^\bullet\mathcal{S}^\bullet$ is constant. In view of 8.1, we may (and do) assume $\mathcal{S}^\bullet$ to be a complex of constant sheaves. Since X is orientable, $\mathcal{D}^\bullet_X[-n]$ is an injective resolution of R_X .

Assume first $\mathcal{S}^\bullet$ to be a sheaf $\mathcal{E}$. Then $\mathfrak{D}^\bullet\mathcal{E}$ is clc by 7.9 . If moreover $\mathcal{E}$ is free with finitely generated stalks, then $\mathcal{E}^{**} = \mathcal{E}$ and therefore 7.10(4) implies that $\mathcal{E} \to \mathfrak{D}^\bullet\mathfrak{D}^\bullet\mathcal{E}$ is an isomorphism. If $\mathcal{E}_x$ is finitely generated, but not necessarily free, then it has a bounded left resolution $\mathcal{C}^\bullet \to \mathcal{E}$ by constant, R-free sheaves with finitely generated stalks. We apply BD_X to $\mathcal{C}^\bullet \to \mathcal{E}$ and get a commutative diagram

$$\begin{array}{ccccccccc}
\cdots \longrightarrow & \mathcal{C}^i & \longrightarrow & \mathcal{C}^{i-1} & \longrightarrow \cdots \longrightarrow & \mathcal{C}^0 & \longrightarrow & \mathcal{E} & \longrightarrow 0 \\
 & \downarrow{\scriptstyle \alpha_i} & & \downarrow{\scriptstyle \alpha_{i-1}} & & \downarrow{\scriptstyle \alpha_0} & & \downarrow{\scriptstyle \beta} & \\
\cdots \longrightarrow & \mathfrak{D}^\bullet\mathfrak{D}^\bullet\mathcal{C}^i & \longrightarrow & \mathfrak{D}^\bullet\mathfrak{D}^\bullet\mathcal{C}^{i-1} & \longrightarrow \cdots \longrightarrow & \mathfrak{D}^\bullet\mathfrak{D}^\bullet\mathcal{C}^0 & \longrightarrow & \mathfrak{D}^\bullet\mathfrak{D}^\bullet\mathcal{E} & \longrightarrow 0
\end{array}$$

where the vertical arrows are defined by BD_X. We have just seen that α_i is a q.i. ($i \geqq 0$). The upper row is exact. Then so is the lower row since $S^\bullet \mapsto \mathcal{D}^\bullet\mathcal{D}^\bullet S^\bullet$ is exact. It is then clear that β is a q.i.

This proves our assertion when $S^\bullet$ is a single degree complex or also (1.10) when $H^\bullet S^\bullet$ is not zero in at most one degree. We then proceed by induction on the length of $H^\bullet S^\bullet$: if b is the greatest integer for which $H^b S^\bullet \neq 0$, we consider the exact sequence

$$(1) \qquad 0 \to \tau_{<b} S^\bullet \to S^\bullet \to \tau^{\geqq b} S^\bullet \to 0$$

from which we get a commutative diagram

$$(2) \qquad \begin{array}{ccccccccc} 0 & \longrightarrow & \tau_{<b} S^\bullet & \longrightarrow & S^\bullet & \longrightarrow & \tau^{\geqq b} S^\bullet & \longrightarrow & 0 \\ & & \downarrow \ell_1 & & \downarrow \ell_2 & & \downarrow \ell_3 & & \\ 0 & \longrightarrow & \mathcal{D}^\bullet\mathcal{D}^\bullet \tau_{<b} S^\bullet & \longrightarrow & \mathcal{D}^\bullet\mathcal{D}^\bullet S^\bullet & \longrightarrow & \mathcal{D}^\bullet\mathcal{D}^\bullet \tau^{\geqq b} S^\bullet & \longrightarrow & 0 \end{array}$$

with exact rows. By induction, ℓ_1 and ℓ_3 are q.i.. Then so is ℓ_2, which concludes the proof of 8.10 in the present case.

b) In the general case, we use the notation of 2.1 and prove by induction on $k \geqq 1$ that $S_k^\bullet \to \mathcal{D}^\bullet\mathcal{D}^\bullet S_k^\bullet$ is a q.i.. For $k = 1$ this follows from a) . So assume it is true for some $k \geqq 1$. Let

$$Z = S_{n-k}\ ,\ U = U_k\ ,\ Y = U_{k+1}\ ,\ j = j_k \text{ and } i = i_k\ .$$

Applying $\mathcal{D}_Y^\bullet\mathcal{D}_Y^\bullet$ to the distinguished triangle

$$(1) \qquad \begin{array}{ccc} j_! j^! S_{k+1}^\bullet & \longrightarrow & S_{k+1}^\bullet \\ & {\scriptstyle [1]} \nwarrow \quad \swarrow & \\ & Ri_* i^* S_{k+1}^\bullet & \end{array}$$

we get a morphism of distinguished triangles which we can write, taking into account the equality $i^*S^\bullet_{k+1} = S^\bullet_k$:

$$\begin{array}{ccccccc} j_!j^!S^\bullet_{k+1} & \longrightarrow & S^\bullet_{k+1} & \longrightarrow & Ri_*S^\bullet_k & \longrightarrow & j_!j^!S^\bullet_{k+1}[1] \\ \downarrow{\scriptstyle\alpha} & & \downarrow{\scriptstyle\beta} & & \downarrow{\scriptstyle\gamma} & & \downarrow{\scriptstyle\alpha[1]} \\ \mathcal{D}^\bullet_Y\mathcal{D}^\bullet_Y j_!j^!S^\bullet_{k+1} & \longrightarrow & \mathcal{D}^\bullet_Y\mathcal{D}^\bullet_Y S^\bullet_{k+1} & \longrightarrow & \mathcal{D}^\bullet_Y\mathcal{D}^\bullet_Y Ri_*S^\bullet_k & \longrightarrow & \mathcal{D}^\bullet_Y\mathcal{D}^\bullet_Y j_!j^!S^\bullet_{k+1}[1] \end{array} \tag{3}$$

We have to prove that β is an isomorphism in $D(X)$. For this, it suffices to show that α and γ are so. Applying 8.8(2) twice, we get

$$\mathcal{D}^\bullet_Y\mathcal{D}^\bullet_Y j_!j^!S^\bullet_{k+1} = j_!\mathcal{D}^\bullet_Z\mathcal{D}^\bullet_Z j^!S^\bullet_{k+1} \quad ;$$

the right hand side being equal to $j_!j^!S^\bullet_{k+1}$ by induction assumption (or a)) we se that α is an isomorphism. By 8.8(3) we have

$$\mathcal{D}^\bullet_Y Ri_*S^\bullet_k = i_!\mathcal{D}^\bullet_U S^\bullet_k \ .$$

Using then 8.8(2), we get

$$\mathcal{D}^\bullet_Y\mathcal{D}^\bullet_Y Ri_*S^\bullet_k = \mathcal{D}^\bullet_Y i_!\mathcal{D}^\bullet_U S^\bullet_k = Ri_*\mathcal{D}^\bullet_U\mathcal{D}^\bullet_U S^\bullet_k \ .$$

Since $\mathcal{D}^\bullet_U\mathcal{D}^\bullet_U S^\bullet_k = S^\bullet_k$ by induction, this shows that γ is an isomorphism in $D(X)$, too, and ends the proof of the theorem.

8.11 COROLLARY . *For any* $S^\bullet \in D^b(\mathfrak{X})$ *and any* $x \in X$, *we have*

$$H^\bullet(S^\bullet_x) = \mathrm{Ext}^\bullet(f^!_x\mathcal{D}^\bullet S^\bullet, R) \ .$$

This follows from the isomorphism 7.7(6), applied to $\mathcal{D}^\bullet S^\bullet$, and from biduality (8.10) .

Remark. The analogue of 8.10 in the complex analytic or algebraic setting, with the notion of constructibility usual in that framework, is proved by J.L. Verdier in [13], see 6.2,p.118 there. Another version under more general assumptions is given in Exp. 10, §2 of [10].

§ 9 PAIRINGS AND POINCARÉ DUALITY IN IC

In this paragraph $(X,\mathfrak{X})$ *is a stratified pseudomanifold of dimension* n .

A. *Some morphisms* .

9.1 LEMMA . *Let* Y *be a topological space,* $A^\bullet \in D(Y)$, $m \in \mathbb{Z}$. *Assume that the natural morphism* $\tau_{\leq m}A^\bullet \to A^\bullet$ *is an isomorphism (in* D(Y)).

Then:

a) *for any* $\mathcal{B}^\bullet \in D(Y)$, *the natural homomorphism*

$$\text{(1)} \qquad \mathrm{Mor}_{D(Y)}(A^\bullet, \tau_{\leq m}\mathcal{B}^\bullet) \longrightarrow \mathrm{Mor}_{D(Y)}(A^\bullet, \mathcal{B}^\bullet)$$

is an isomorphism.

b) *Let* $U \subset Y$ *be open and* $i : U \to Y$ *be the inclusion map. Then for any* $\mathcal{B}'^\bullet \in D(U)$ *the natural homomorphism*

$$\text{(2)} \qquad \mathrm{Mor}_{D(Y)}(A^\bullet, \tau_{\leq m}Ri_*\mathcal{B}'^\bullet) \longrightarrow \mathrm{Mor}_{D(U)}(i^*A^\bullet, \mathcal{B}'^\bullet)$$

is an isomorphism.

Proof. Recall (5.12) that a morphism from $A^\bullet$ to $\mathcal{B}^\bullet$ in D(Y) is represented by a diagram

$$\text{(3)} \qquad A^\bullet \xleftarrow{\sim} C^\bullet \longrightarrow \mathcal{B}^\bullet$$

in K(Y). Since $\tau_{\leq m}A^\bullet \to A^\bullet$ is a q.i., so is $\tau_{\leq m}C^\bullet \to C^\bullet$, and (3) is equivalent to the diagram

$$\text{(4)} \qquad A^\bullet \xleftarrow{\sim} \tau_{\leq m}(C^\bullet) \longrightarrow \mathcal{B}^\bullet .$$

But $\tau_{\leq m}(C^\bullet) \to \mathcal{B}^\bullet$ factors through $\tau_{\leq m}\mathcal{B}^\bullet$. Thus (1) is surjective. Injectivity can be checked in a similar way. Indeed, given a commutative diagram

(5) 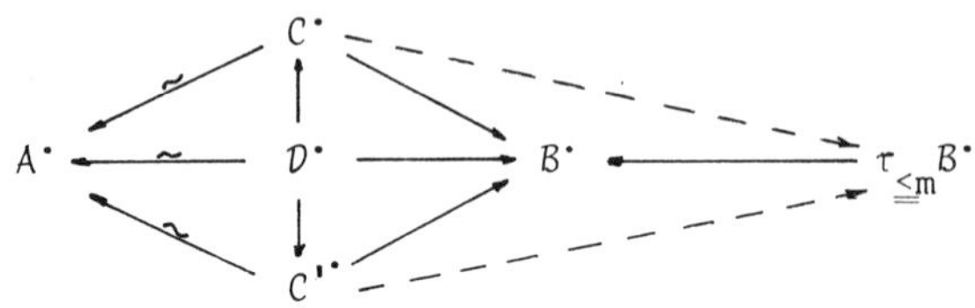

In K(Y), we can replace $\mathcal{D}^\bullet$ by $\tau_{\leqq m}\mathcal{D}^\bullet$ and factor $\tau_{\leqq m}\mathcal{D}^\bullet \to \mathcal{B}^\bullet$ through $\tau_{\leqq m}\mathcal{B}^\bullet$. It is readily checked that the resulting diagram remains commutative in K(Y). This proves (a). Now (b) follows from the adjunction isomorphism

$$(6) \qquad \mathrm{Mor}_{D(Y)}(A^\bullet, Ri_*\mathcal{B}'^\bullet) = \mathrm{Mor}_{D(U)}(i^*A^\bullet, \mathcal{B}'^\bullet)$$

and (a), applied to $\mathcal{B}^\bullet = Ri_*\mathcal{B}'^\bullet$.

9.2 PROPOSITION . *Let* $f_2 : E \to F$ *be a morphism of local systems on* U_2 *and* p,q *be perversities. Assume that* $p \leqq q$ (i.e. $p(k) \leqq q(k)$ *for every* $k \geqq 2$) . *Then* f_2 *extends in a unique way to a morphism* $f : P_p^\bullet(E) \to P_q^\bullet(F)$ *in* D(X) .

Proof . Let $L^\bullet = P_p^\bullet(E)$, $M^\bullet = P_q^\bullet(F)$ and let $k \geqq 2$. It is enough to check that any morphism $f_k : L_k^\bullet \to M_k^\bullet$ in $D(U_k)$ extends uniquely to a morphism $f_{k+1} : L_{k+1}^\bullet \to M_{k+1}^\bullet$ in $D(U_{k+1})$. By definition $M_{k+1}^\bullet = \tau_{\leqq q(k)} Ri_{k*} M_k^\bullet$, and $\tau_{\leqq q(k)} L_{k+1}^\bullet = L_{k+1}^\bullet$. Thus by 9.1(b)

$$\mathrm{Mor}_{D(U_{k+1})}(L_{k+1}^\bullet, \tau_{\leqq q(k)} Ri_{k*}M_k) = \mathrm{Mor}_{D(U_k)}(L_k^\bullet, M_k^\bullet)$$

and the result follows.

9.3 LEMMA . *For any* $k \geqq 2$, *the attachment map*

(1) $$\alpha : \mathcal{D}^\bullet_{U_{k+1}} \longrightarrow Ri_{k*}\mathcal{D}^\bullet_{U_k}$$

is a quasi-isomorphism up to $k-2-n$.

Proof. Let $S^\bullet = \mathcal{D}^\bullet_X$. Then 8.2 implies

$$j_k^! S^\bullet_{k+1} = j_k^! \mathcal{D}^\bullet_{U_{k+1}} = \mathcal{D}^\bullet_{S_{n-k}} j_k^* R_X = \mathcal{D}^\bullet_{S_{n-k}} .$$

But $\mathcal{D}^i_{S_{n-k}} = 0$ for $i < -\dim S_{n-k} = k-n$. Thus $H^i(j_k^! S^\bullet_{k+1}) = 0$ for $i < k-n$. The result follows now from the long exact sequence 1.8(7):

$$\cdots \longrightarrow \mathcal{H}^i(j_{k!} j_k^! S^\bullet_{k+1}) \longrightarrow \mathcal{H}^i(S^\bullet_{k+1}) \longrightarrow \mathcal{H}^i(Ri_{k*} S^\bullet_k) \longrightarrow \cdots$$

9.4 PROPOSITION . *Consider on* U_2 *a local system* E *, the orientation sheaf* $\mathcal{O}$ *and a morphism* $f_2 : E \to \mathcal{O}$ *. Let* p *be a perversity. Then there exists a unique morphism* $f : P^\bullet_p(E) \to \mathcal{D}^\bullet_X[-n]$ *in* $D(X)$ *which extends* f_2 .

In view of 9.3 , the proof is essentially the same as for 9.2 : Let $L^\bullet = P^\bullet_p(E), M^\bullet = \mathcal{D}^\bullet_X[-n]$. For $k \geq 2$ we have $p(k) \leq k-2$; using successively 9.1(a), 9.3 and 9.1(b) we find

$$\mathrm{Mor}_{D(U_{k+1})}(L^\bullet_{k+1}, M^\bullet_{k+1}) = \mathrm{Mor}_{D(U_{k+1})}(L^\bullet_{k+1}, \tau_{\leq k-2} M^\bullet_{k+1}) =$$

$$= \mathrm{Mor}_{D(U_{k+1})}(L^\bullet_{k+1}, \tau_{\leq k-2} Ri_{k*} M^\bullet_k) = \mathrm{Mor}_{D(U_k)}(L^\bullet_k, M^\bullet_k) .$$

By induction on k, this proves the proposition.

B. *Poincaré Duality*.

9.5 LEMMA. *Let* Y *be a topological space with a filtration* $\mathfrak{Y}$ *by closed subspaces* $Y_m = Y \supset Y_{m-1} \supset \ldots \supset Y_{-1} = \emptyset$. *Assume that each stratum* $Y_i - Y_{i-1}$ *is a manifold of dimension* i *or is empty* $(0 \leq i \leq m)$. *Let* $\ell \in \mathbb{Z}$ *and* $S^\bullet \in DGS(Y)$ *be such that* $H^j(S^\bullet_y) = 0$ *if* $y \in Y_k - Y_{k-1}$ *and* $j > \ell - k$. *Then* $\mathbb{H}^j_c(Y;S^\bullet) = 0$ *for* $j > \ell$.

Proof. We use induction on m. The lemma is clearly true for $m = 0$. Let now $m > 0$ and assume that the result holds for $m - 1$. We have a long exact sequence

$$(1) \qquad \ldots \longrightarrow \mathbb{H}^j_c(Y-Y_{m-1};S^\bullet) \longrightarrow \mathbb{H}^j_c(Y;S^\bullet) \longrightarrow \mathbb{H}^j_c(Y_{m-1};S^\bullet) \longrightarrow \ldots$$

By induction hypothesis $\mathbb{H}^j_c(Y_{m-1};S^\bullet) = 0$ for $j > \ell$. Since $Y - Y_{m-1}$ is a manifold of dimension m (or is empty) and $H^i(S^\bullet|_{Y-Y_{m-1}}) = 0$ for $i > \ell - m$, it follows from 6.8 that $S^\bullet|_{Y-Y_{m-1}}$ has a c-soft resolution $A^\bullet$ with $A^j = 0$ for $j > \ell$. Therefore $\mathbb{H}^j_c(Y-Y_{m-1};S^\bullet) = 0$ for $j > \ell$. The exact sequence (1) gives then $\mathbb{H}^j_c(Y;S^\bullet) = 0$ for $j > \ell$.

9.6 COROLLARY. *Let* E *be a local system on* U_2 *and let* p *be a perversity. Then*

$$\mathbb{H}^j_c(X;P^\bullet_p(E)) = 0 \text{ for } j > n \text{ and } \mathbb{H}^n_c(X;P^\bullet_p(E)) = H^n_c(U_2;E).$$

Let $x \in S_k \subset X_{n-2}$. By definition of $P^\bullet_p(E)$ we have $H^j(P^\bullet_p(E)_x) = 0$ for $j > p(n-k)$, in particular for $j > n - k - 2$. By 9.5 we have therefore

$$\mathbb{H}^j_c(X_{n-2};P^\bullet_p(E)) = 0 \text{ for } j > n - 2.$$

From the long exact sequence:

$$\ldots \longrightarrow H_c^j(U_2;E) \longrightarrow \mathbb{H}_c^j(X;P_p^\bullet(E)) \longrightarrow \mathbb{H}_c^j(X_{n-2};P_p^\bullet(E)) \longrightarrow \ldots$$

we deduce then that

$$\mathbb{H}_c^j(X;P_p^\bullet(E)) = H_c^j(U_2;E) \qquad \text{for } j \geqq n ,$$

and it is well known that $H_c^j(U_2;E) = 0$ for $j > n$.

9.7 Assume now that R is a field. Let $V^* = \mathrm{Hom}(V,R)$ denote the dual of a vector space V over R . If $S^\bullet \in \mathrm{DGS}(X)$, 7.7(2) becomes

(1) $$\mathbb{H}^i(U;\mathbb{D}_X^\bullet S^\bullet) = \mathbb{H}_c^{-i}(U;S^\bullet)^* \quad (U \text{ open in } X \text{ , } i \in \mathbb{Z})$$

Suppose that $S^\bullet$ is $\mathfrak{X}$ - clc . Letting U run over a fundamental system of distinguished neighborhoods of $x \in X$, we get then by 3.10

(2) $$H^i(\mathbb{D}_X^\bullet S^\bullet)_x = H^{-i}(f_x^! S^\bullet)^* \quad (i \in \mathbb{Z} \text{ , } x \in X)$$

Suppose moreover that $S^\bullet$ is $\mathfrak{X}$ - cc . Then, similarly, 8.11 gives

(3) $$H^i(S_x^\bullet) = H^{-i}(f_x^! \mathbb{D}_X^\bullet S^\bullet)^* \quad (x \in X \text{ , } i \in \mathbb{Z})$$

Replacing $\mathbb{D}_X^\bullet S^\bullet$ by $\mathbb{D}_X^\bullet S^\bullet[-n]$ and modifying the indices in a suitable way, we get from these formulae :

(1') $$\mathbb{H}^i(U;\mathbb{D}_X^\bullet S^\bullet[-n]) = \mathbb{H}_c^{n-i}(U;S^\bullet)^* \quad (i \in \mathbb{Z}, \text{ U open in X})$$

(2') $$H^i(\mathbb{D}_X^\bullet S^\bullet[-n]_x) = H^{n-i}(f_x^! S^\bullet)^* \quad (i \in \mathbb{Z}, \ x \in X)$$

(3') $$H^i(S_x^\bullet) = H^{n-i}(f_x^! \mathbb{D}_X^\bullet S^\bullet[-n])^* \quad (i \in \mathbb{Z}, \ x \in X)$$

(where we write $\mathbb{D}_X^\bullet S^\bullet[-n]$ for $(\mathbb{D}_X^\bullet S^\bullet)[-n]$). We can now prove

9.8 THEOREM . *Let* R *be a field and let* p,q *be complementary perversities. If* $\mathcal{E}$ *is a local system on* U_2, *then there exists a unique isomorphism*

$$(1) \qquad \mathbb{D}_X^{\bullet} P_p^{\bullet}(\mathcal{E})[-n] = P_q^{\bullet}(\mathbb{D}_{U_2}^{\bullet} \mathcal{E}[-n])$$

in D(X) *which extends the identity map of* $\mathbb{D}_{U_2}^{\bullet} \mathcal{E}[-n]$.

Proof . As R is a field, $\mathcal{E}$ is locally free. By 7.10(4) we have then $\mathbb{D}_{U_2}^{\bullet} \mathcal{E}[-n] = \mathcal{E}^* \otimes \mathcal{O}$, where $\mathcal{E}^* = \mathrm{Hom}(\mathcal{E}, R_{U_2})$ and $\mathcal{O}$ is the orientation sheaf. In particular $\mathbb{D}_{U_2}^{\bullet} \mathcal{E}^*[-n]$ is again a local system, so that the right hand side of (1) makes sense.

Let $S^{\bullet} = P_p^{\bullet}(\mathcal{E})$. We check first that $\mathbb{D}_X^{\bullet} S^{\bullet}[-n]$ satisfies $(AX2)_{\mathfrak{X}, \mathcal{E}^* \otimes \mathcal{O}, q}$ weakened along the lines of 2.7(b), that is, with the condition $(\mathbb{D}_X^{\bullet} S^{\bullet}[-n])^i = 0$ for $i < 0$ replaced by $\mathcal{H}^i(\mathbb{D}_X^{\bullet} S^{\bullet}[-n]) = 0$ for $i < 0$.

(a) By 8.7 , $\mathbb{D}_X^{\bullet} S^{\bullet}[-n]$ is bounded and $\mathfrak{X}$ - clc , and $\mathbb{D}_X^{\bullet} S^{\bullet}[-n]|_{U_2} = \mathbb{D}_{U_2}^{\bullet} \mathcal{E}[-n]$ since $S^{\bullet}|_{U_2} = \mathcal{E}$.

Let U be a distinguished neighborhood of $x \in X$ and let $i < 0$. Using successively 9.7(2'),3.10 and 9.6 , we get

$$H^i(\mathbb{D}_X^{\bullet} S^{\bullet}[-n]_x) = H^{n-i}(f_x^! S^{\bullet})^* = \mathbb{H}_c^{n-i}(U; S^{\bullet})^* = 0$$

Thus $H^i(\mathbb{D}_X^{\bullet} S^{\bullet}[-n] = 0$ for $i < 0$.

(b) By 9.7(2'), $H^i(\mathbb{D}_X^{\bullet} S^{\bullet}[-n]_x) = H^{n-i}(f_x^! S^{\bullet})^*$. Therefore

$$\mathrm{supp}\, \mathcal{H}^i(\mathbb{D}_X^{\bullet} S^{\bullet}[-n]) = \{x \in X \mid H^{n-i}(f_x^! S^{\bullet}) \neq 0\}$$

As $S^{\bullet}$ satisfies condition (c) of $(AX2)_{\mathfrak{X}, \mathcal{E}, p}$ we get for $i > 0$

$$\dim \mathrm{supp}\, \mathcal{H}^i(\mathbb{D}_X^{\bullet} S^{\bullet}[-n]) \leq n - q^{-1}(n-(n-i)) = n - q^{-1}(i) .$$

Condition (c) follows in a similar way from condition (b) of

$(AX2)_{\mathbf{X},E,p}$ for $S^\bullet$.

Thus $\mathbf{D}_X^\bullet P_p^\bullet(E)[-n]$ satisfies $(AX2)_{\mathbf{X},E^* \otimes \mathcal{O},q}$, and so does of course $P_q^\bullet(E^* \otimes \mathcal{O})$. The existence of the isomorphism (1) follows then from 4.10 , and its uniqueness from 9.2 .

9.9 COROLLARY . *We have a natural isomorphism*

$$(1)\qquad I_qH^i(X;E^* \otimes \mathcal{O}) = I_pH_c^{n-i}(X;E)^* . \quad (i \in \mathbb{Z}) .$$

Proof: We must show that

$$(2)\qquad \mathbb{H}^i(X;P_q^\bullet(E^* \otimes \mathcal{O})) = \mathbb{H}_c^{n-i}(X;P_p^\bullet(E))^* .$$

Since $E^* \otimes \mathcal{O} = \mathbf{D}_{U_2}^\bullet E[-n]$, this is equivalent by 9.8 to

$$(3)\qquad \mathbb{H}^i(X;\mathbf{D}_X^\bullet P_p^\bullet(E)[-n]) = \mathbb{H}_c^{n-i}(X;P_p^\bullet(E))^* ,$$

which can be deduced from 9.7(1) by setting $S^\bullet = P_p^\bullet(E), U = X$ and shifting by n .

9.10 *Examples*. In this section X is *normal* .

i) Take $E = \mathcal{O}$, $p = t$, $q = 0$ (t is the top perversity) . By 2.8 and 2.12 , we have

$$I_qH^i(X;\mathcal{O}^* \otimes \mathcal{O}) = I_oH^i(X;R) = H^i(X;R)$$

$$I_pH_c^{n-i}(X;\mathcal{O}) = I_tH_c^{n-i}(X;\mathcal{O}) = H_{i,c}(X,R)$$

and 9.9 gives

$$H^i(X;R) = H_{i,c}(X;R)^* .$$

ii) Take $E = R$, $p = 0$, $q = t$. Then by 2.12 and 2.8 , we get

$$I_qH^i(X;R^* \otimes \mathcal{O}) = I_tH^i(X;\mathcal{O}) = H_{n-i}(X;R)$$

$$I_pH_c^{n-i}(X;R) = I_0H_c^{n-i}(X;R) = H_c^{n-i}(X;R)$$

and 9.9 yields

$$H_{n-i}(X;R) = H_c^{n-i}(X;R)^* .$$

9.11 Let R be a field. Suppose that $\mathfrak{X}$ has only even-codimensional strata and that U_2 is orientable (i.e. the orientation sheaf on U_2 is isomorphic to the constant sheaf R). Let m be the middle perversity. If $S^\bullet \in DGS(X)$, then $S^\bullet$ satisfies $(AX2)_{R,m}$ if and only if $\mathbb{D}_X^\bullet S^\bullet[-n]$ does so. We are thus led to consider the set of conditions :

AX3 .

(3a) $S^\bullet$ is bounded, $\mathcal{H}^i(S^\bullet) = 0$ for $i < 0$, $S^\bullet$ is the constant sheaf R on the dense stratum of some topological stratification of X, and $S^\bullet$ is $\mathfrak{X}$ - clc for some topological stratification $\mathfrak{X}$ of X .

(3b) dim supp $\mathcal{H}^i(S^\bullet) \leqq n - 2i - 2$ for $i > 0$.

(3c) $\mathbb{D}_X^\bullet S^\bullet[-n]$ is isomorphic to $S^\bullet$ in $D^b(X)$.

9.12 PROPOSITION . *Let* $S^\bullet \in DGS(X)$ *be bounded. Then in the situation of* 9.11 *the following conditions are equivalent:*

a) $S^\bullet$ *satisfies* AX3 .

b) $\mathcal{H}^i(S^\bullet) = 0$ *for* $i < 0$ *and* $\tau_{\geqq 0}S^\bullet$ *satisfies* $(AX2)_{R,m}$.

Proof : If (b) holds, then 9.8 shows that $S^\bullet$ satisfies AX3 . Conversely, if $S^\bullet$ satisfies AX3 , then (2a) and (2b) certainly hold for $\tau_{\geqq 0}S^\bullet$, and as in the proof of 9.5 we find that (2c) holds because $\mathbb{D}_X^\bullet S^\bullet[-n]$ satisfies (2b). Notice that the passage from condition (2b) for $\mathbb{D}_X^\bullet S^\bullet[-n]$ to (2c) for $S^\bullet$ uses only 9.7(2'), which requires only $S^\bullet$ to be $\mathfrak{X}$ - clc .

C. *Pairings*.

9.13 LEMMA. *Let* Y *be a topological space and let* $A^\cdot, B^\cdot \in D(Y)$ *satisfy* $H^i(A^\cdot) = 0$ *for* $i > 0$, $H^i(B^\cdot) = 0$ *for* $i < 0$. *Then the natural homomorphism*

$$\mathrm{Mor}_{D(Y)}(A^\cdot, B^\cdot) \longrightarrow \mathrm{Hom}_{Sh(Y)}(H^0(A^\cdot), H^0(B^\cdot))$$

is an isomorphism.

Proof. We can assume that $B^i = 0$ for $i < 0$. The morphisms from $A^\cdot$ to $B^\cdot$ in D(Y) are represented by diagrams in K(Y)

(1) $$A^\cdot \xleftarrow{s} C^\cdot \xrightarrow{f} B^\cdot$$

with s a q.i.. The lemma is a direct consequence of the following observations:

a) If in (1) we replace $C^\cdot$ by $\tau_{\leq 0}C^\cdot$, the resulting diagram represents the same morphism in D(Y).

b) If in (1) we have $C^i = 0$ for $i > 0$, then the natural homomorphism

$$\mathrm{Mor}_{K(Y)}(C^\cdot, B^\cdot) \to \mathrm{Hom}_{Sh(Y)}(H^0(C^\cdot), H^0(B^\cdot)) = \mathrm{Hom}_{Sh(Y)}(H^0(A^\cdot), H^0(B^\cdot))$$

is an isomorphism.

c) Given $\phi, \phi' \in \mathrm{Mor}_{D(Y)}(A^\cdot, B^\cdot)$, there exist $C^\cdot$, a q.i. $s : C^\cdot \to A^\cdot$ and morphisms $f, f' : C^\cdot \to B^\cdot$ in K(Y) such that (s,f) and (s,f') represent ϕ and ϕ' respectively. (In other words ϕ and ϕ' have a common denominator). This follows from 5.8.

Remark: Under the same hypothesis, the natural homomorphism

$$\mathrm{Mor}_{D(Y)}(B^\cdot, A^\cdot) \longrightarrow \mathrm{Hom}_{Sh(Y)}(H^0(B^\cdot), H^0(A^\cdot))$$

need not be an isomorphism.

9.14 PROPOSITION. *Let* $\mu_2 : E \otimes F \to G$ *be a pairing of local systems on* U_2 *and let* p,q,r *be perversities such that* $p(k) + q(k) \leq r(k)$

$(2 \leq k \leq n)$. *Then there exists in* $D^b(X)$ *a unique morphism*

$\mu : P^\bullet_p(E) \overset{L}{\otimes} P^\bullet_q(F) \to P^\bullet_r(G)$ *which coincides with* μ_2 *over* U_2 . *Moreover, if* G *is the orientation sheaf, then there exists a unique morphism*

$\mu' : P^\bullet_p(E) \overset{L}{\otimes} P^\bullet_q(F) \to \mathcal{D}^\bullet_X[-n]$ *which coincides with* μ_2 *over* U_2 .

Proof. Notice first that $H^i(E \overset{L}{\otimes} F) = 0$ for $i > 0$ and $H^0(E \overset{L}{\otimes} F) = E \otimes F$. By 9.13 we can therefore consider μ_2 as a morphism from $E \overset{L}{\otimes} F$ to G in $D(U_2)$ so that the statement of the proposition makes sense. Let

$$L^\bullet = P^\bullet_p(E),\ M^\bullet = P^\bullet_q(F),\ A^\bullet = L^\bullet \overset{L}{\otimes} M^\bullet .$$

We claim

(1) $$H^i(A^\bullet_{k+1}) = 0 \text{ for } i > r(k) \quad (k \geq 2) .$$

Indeed, $A^\bullet_{k+1}$ is quasi-isomorphic to $L^\bullet_{k+1} \overset{L}{\otimes} M^\bullet_{k+1}$. To compute the latter, we can first replace $L^\bullet_{k+1}$ and $M^\bullet_{k+1}$ by $\tau_{\leq p(k)} L^\bullet_{k+1}$ and $\tau_{\leq q(k)} M^\bullet_{k+1}$ respectively. We can then choose a flat resolution $F^\bullet \to \tau_{\leq p(k)} L^\bullet_{k+1}$ with $F^i = 0$ for $i > p(k)$. Then $A^\bullet_{k+1}$ is quasi-isomorphic to $F^\bullet \otimes \tau_{\leq q(k)} M^\bullet_{k+1}$ and $(F^\bullet \otimes \tau_{\leq q(k)} M^\bullet_{k+1})^i = 0$ for $i > r(k)$.

We can now conclude as in 9.2 and 9.4 . Let $N^\bullet$ be one of the complexes $P^\bullet_r(G)$ or, if $G = \mathcal{O}, \mathcal{D}^\bullet_X[-n]$. Using 9.1 and 9.3, we see that for any $A^\bullet \in D(X)$ satisfying (1), any morphism $f_k : A^\bullet_k \to N^\bullet_k$ in $D(U_k)$ can be extended in a unique way to a morphism $f_{k+1} : A^\bullet_{k+1} \to N^\bullet_{k+1}$ in $D(U_{k+1})$. Using induction on $k \geq 2$, we get the existence and uniqueness of the required extension of μ_2 .

9.15 The pairing $\mu : P^\bullet_p(E) \overset{L}{\otimes} P^\bullet_q(F) \to P^\bullet_r(G)$ of 9.14 gives rise to various pairings in hypercohomology. In particular we get a map

(1) $$\mathbb{H}^i_c(X;P^\bullet_p(E)) \otimes \mathbb{H}^j(X;P^\bullet_q(F)) \to \mathbb{H}^{i+j}_c(X;P^\bullet_r(G)) .$$

Let $E^* = Hom(E,R)$. We can take $F = E^* \otimes \mathcal{O}$, $G = \mathcal{O}$, and for μ_2 the canonical pairing $E \otimes (E^* \otimes \mathcal{O}) \to \mathcal{O}$. By 9.3 we know that $\mathbf{H}^n_c(X;P^\cdot_r(\mathcal{O})) = H^n_c(U_2;\mathcal{O})$. When $i + j = n$ we can compose in this case (1) with the canonical map from $H^n_c(U_2,\mathcal{O})$ to R . We get then a natural pairing

$$(2) \qquad I_pH^{n-j}_c(X;E) \otimes I_qH^j(X;E^* \otimes \mathcal{O}) \to R .$$

9.16 PROPOSITION . *If* R *is a field and* p,q *are complementary perversities, then* 9.15(2) *coincides with the pairing*

$$(1) \qquad I_pH^{n-j}_c(X;E) \otimes I_qH^j(X;E^* \otimes \mathcal{O}) \to R$$

induced by the isomorphism 9.9 .

Proof . 9.9 is defined by means of the isomorphism 9.8(1)

$$(2) \qquad P^\cdot_q(E^* \otimes \mathcal{O}) \xrightarrow{\sim} Hom^\cdot(P^\cdot_p(E),\mathcal{D}^\cdot_X[-n]) .$$

The latter also yields a morphism

$$(3) \qquad \mu'' : P^\cdot_p(E) \overset{L}{\otimes} P^\cdot_q(E^* \otimes \mathcal{O}) \to \mathcal{D}^\cdot_X[-n] .$$

Now (3) induces a pairing of hypercohomology groups

$$(4) \qquad \mathbf{H}^{n-j}_c(X;P^\cdot_p(E)) \otimes H^j(X;P^\cdot_q(E^* \otimes \mathcal{O})) \to \mathbf{H}^n_c(X;\mathcal{D}^\cdot_X[-n])$$

which, composed with the natural map from $\mathbf{H}^n_c(X;\mathcal{D}^\cdot_X[-n])$ to R , gives (1) back.

On the other hand, 9.15(2) comes from the pairing

$$(5) \qquad \mu : P^\cdot_p(E) \overset{L}{\otimes} P^\cdot_q(E^* \otimes \mathcal{O}) \to P^\cdot_t(\mathcal{O})$$

provided by 9.14 . Let $\beta : P^\cdot_t(\mathcal{O}) \to \mathcal{D}^\cdot_X[-n]$ be the morphism given by

9.4. It is clear that μ'' restricts over U_2 to the natural pairing $\mu_2 : \mathcal{E} \otimes (\mathcal{E}^* \otimes \mathcal{O}) \to \mathcal{O}$. By the uniqueness statement in 9.14 , we have therefore $\mu'' = \beta \circ \mu$. This implies in particular the equality of (1) and 9.15(2) .

§ 10 SOME FORMULAE IN DERIVED CATEGORIES OF SHEAVES

In this section we collect and prove a number of general identities pertaining to derived categories of sheaves and the duality functor. Together with the previous §§ , this will include all the identities in [6] (except for 1.13.(13), not always true). A number of them have already appeared, at least in special cases, in the previous sections but the discussion was often limited to what was needed at the moment.

A. *Continuous maps*

In this part, the spaces satisfy from 10.2 on our usual standing assumptions (locally compact, locally completely paracompact (1.17), finite cohomological dimension over R*) and* $f : X \to Y$ *is a continuous map.*

10.1 PROPOSITION . *We have in* D(X)

(1) $$f^*(A^\cdot \overset{L}{\otimes} B^\cdot) = f^*A^\cdot \overset{L}{\otimes} f^*B^\cdot \qquad (A^\cdot, B^\cdot \in D(Y)) .$$

If we replace $B^\cdot$ by a flat left resolution we are reduced to prove in Sh(X)

(2) $$f^*(A \otimes B) = f^*A \otimes f^*B , \qquad (A,B \in Sh(Y)) ,$$

which follows immediately from the relation $(A \otimes B)_x = A_x \otimes B_x$.

10.2 PROPOSITION . Let $A^\cdot, B^\cdot, C^\cdot \in D(X)$. Then

(1) $$RHom^\cdot(A^\cdot \overset{L}{\otimes} B^\cdot, C^\cdot) = RHom^\cdot(A^\cdot, RHom^\cdot(B^\cdot, C^\cdot)) .$$

Proof. We note first that we have

(2) $$Hom(A \otimes B, C) = Hom(A, Hom(B,C) \qquad (A,B,C \in Sh(X)) ,$$

as follows from the corresponding identity for modules, applied to sections over open subsets. From this, it follows :

(3) *If* $\mathcal{B}$ *is flat and* $\mathcal{C}$ *injective, then* $\mathcal{H}om(\mathcal{B},\mathcal{C})$ *is injective*

because (2) shows that $\mathcal{A} \mapsto \mathcal{H}om(\mathcal{A},\mathcal{H}om(\mathcal{B},\mathcal{C})$ is an exact functor.

Now in order to prove (1) we may assume that $\mathcal{B}^\bullet$ is flat and $\mathcal{C}^\bullet$ injective, and then (1) becomes

(4) $$\mathcal{H}om^\bullet(\mathcal{A}^\bullet \otimes \mathcal{B}^\bullet,\mathcal{C}^\bullet) = \mathcal{H}om^\bullet(\mathcal{A}^\bullet,\mathcal{H}om^\bullet(\mathcal{B}^\bullet,\mathcal{C}^\bullet)) ,$$

but this follows from (2) .

10.3 PROPOSITION . *In* D(Y), *we have the equalities*

(1) $$Rf_*R\mathcal{H}om^\bullet(f^*\mathcal{A}^\bullet,\mathcal{B}^\bullet) = R\mathcal{H}om^\bullet(\mathcal{A}^\bullet,Rf_*\mathcal{B}^\bullet) \quad (\mathcal{A}^\bullet \in D(Y),\mathcal{B}^\bullet \in D(X)).$$

(2) $$\mathrm{Mor}_{D(X)}(f^*\mathcal{A}^\bullet,\mathcal{B}^\bullet) = \mathrm{Mor}_{D(Y)}(\mathcal{A}^\bullet,Rf_*\mathcal{B}^\bullet)$$

Proof . We may assume $\mathcal{B}^\bullet$ to be injective. Then so is $f_*\mathcal{B}^\bullet$, and $\mathcal{H}om^\bullet(f^*\mathcal{A}^\bullet,\mathcal{B}^\bullet)$ is flabby. To prove (1) we are then reduced to showing :

(3) $$f_*\mathcal{H}om^\bullet(f^*\mathcal{A}^\bullet,\mathcal{B}^\bullet) = \mathcal{H}om^\bullet(\mathcal{A}^\bullet,f_*\mathcal{B}^\bullet) .$$

For this, it is enough to show the following equality in Sh(Y)

(4) $$f_*\mathcal{H}om(f^*\mathcal{A},\mathcal{B}) = \mathcal{H}om(\mathcal{A},f_*\mathcal{B}) \quad (\mathcal{A} \in Sh(Y),\mathcal{B} \in Sh(X)) .$$

But this is just another way to write the standard adjunction formula (VI,1.4)

(5) $$\mathrm{Hom}(f^*\mathcal{A},\mathcal{B}) = \mathrm{Hom}(\mathcal{A},f_*\mathcal{B}) \quad (\mathcal{A} \in Sh(Y),\mathcal{B} \in Sh(X))$$

for Y and all its open subsets. This proves (1). Now (2) follows from (1) and 5.17(3) .

10.4 As a counterpart to 10.3 , we recall the adjointness properties of $f^!$ and $f_!$. The latter is the functor direct image with proper supports (7.11). Given a flat c-soft sheaf K on X , we have defined a functor $f^!_K$: Sh(Y) $\rightarrow$ Sh(X) which satisfies

(1) $$\mathrm{Hom}(f_!(A \otimes K), B) = \mathrm{Hom}(A, f^!_K B) \quad (A \in \mathrm{Sh}(X), B \in \mathrm{Sh}(Y))$$

Furthermore, $A \mapsto f_!(A \otimes K)$ is exact and, if B is injective, then so is $f^!_K B$ (7.14). In the derived categories, (1) leads to the Verdier duality (7.17)

(2) $$R\mathcal{H}om^\bullet(Rf_! A^\bullet, B^\bullet) = Rf_* R\mathcal{H}om^\bullet(A^\bullet, f^! B^\bullet) \quad ,(A^\bullet \in D(X), B^\bullet \in D(Y)),$$

and (see 7.19(1)) to:

(3) $$\mathrm{Mor}_{D(Y)}(Rf_! A^\bullet, B^\bullet) = \mathrm{Mor}_{D(X)}(A^\bullet, f^! B^\bullet) \ .$$

where $f^! B^\bullet$ is defined as $f^!_K . J^\bullet$, with $J^\bullet$ an injective resolution of $B^\bullet$ and $K^\bullet$ a flat c-soft resolution of R_X .

10.5 LEMMA . (i) *Let* $A^\bullet \in \mathrm{DGS}(X)$. *For* $V \subset Y$ *open, let* $\psi(V) = \{C \cap f^{-1}V | C \subset X \text{ compact}\}$ *and* $f_c A^\bullet$ *be the* DGS *on* Y *associated to the presheaf* $V \mapsto \Gamma_{\psi(V)}(f^{-1}V; A^\bullet)$. *Then* $f_c A^\bullet = f_! A^\bullet$.

(ii) *The functor* $f_!$ *preserves c-softness* .

Proof : (a) We have clearly $\psi(V) \subset \Phi(V)$ whence an inclusion

(1) $$\Gamma_{\psi(V)}(f^{-1}V; A^\bullet) \to \Gamma_{\Phi(V)}(f^{-1}V; A^\bullet)$$

which induces an injective homomorphism

(2) $$\mu : f_c A^\bullet \to f_! A^\bullet \ .$$

Let V' be relatively compact, open with closure in V . If $C \in \Phi(V)$, then $C \cap f^{-1}V'$ has a compact closure, hence belongs to $\psi(V')$. Therefor

$$\mathrm{Im}(\Gamma_{\psi(V')}(f^{-1}V'; A^\bullet) \to \Gamma_{\Phi(V')}(f^{-1}V; A^\bullet)) \supset$$
$$\supset \mathrm{Im}(\Gamma_{\Phi(V)}(f^{-1}V; A^\bullet) \to \Gamma_{\Phi(V')}(f^{-1}V'; A^\bullet))$$

and μ is surjective. This proves (i). The equality

$$(3) \qquad (R^i f_c \mathcal{B})_y \cong H^i_c(f^{-1}y;\mathcal{B}) \qquad (y \in Y; i \in \mathbb{Z}; \mathcal{B} \in \mathrm{Sh}(X)),$$

which follows from [4:IV,4.2] can then also be derived from 7.12 .

(b) We now prove (ii). Let $\mathcal{A} \in \mathrm{Sh}(X)$ be c-soft. We have to show that $f_! \mathcal{A}$ is c-soft. This amounts to proving that $H^1_c(V;f_!\mathcal{A}) = 0$ for all V open in Y . Consider the spectral sequence (E_r) of f for cohomology with compact supports. It abuts to $H^\bullet_c(X;\mathcal{A})$ and we have

$$(4) \qquad E_2^{p,q} = H^p(V;R^i f_c \mathcal{A}) \qquad (p,q \in \mathbb{Z}).$$

Since $\mathcal{A}|f^{-1}V$ is c-soft we have, in view of (3), $R^i f_c \mathcal{A} = 0$ for $i \geqq 1$, hence, taking (a) into account, we get

$$(5) \qquad H^i_c(f^{-1}V;\mathcal{A}) = H^i_c(V;f_c\mathcal{A}) = H^i_c(V;f_!\mathcal{A}) \qquad (i \in \mathbb{Z}) .$$

But the first term is zero for $i \geqq 1$ since $\mathcal{A}$ is c-soft. This proves (ii) .

Remark . I had proved originally that $Rf_c\mathcal{A}^\bullet = Rf_!\mathcal{A}^\bullet$. Spaltenstein pointed out that μ is in fact an isomorphism in DGS(Y).

10.6 THEOREM . *Let* $g : Y \to Z$ *be a continuous map.*

(i) $\qquad (g \circ f)^* = f^* \circ g^*$, $R(g \circ f)_* = Rg_* \circ Rf_*$,

(ii) $\qquad (g \circ f)^! = f^! \circ g^!$, $R(g \circ f)_! = Rg_! \circ Rf_!$.

It is clear that the first equality of (i) and $(f \circ g)_* = f_* \circ g_*$ are true at the sheaf level. Then (i) follows from the fact that f^* is exact and f_* preserves injective sheaves.

(ii) We first note that $(f \circ g)_! = f_! \circ g_!$ at the sheaf level. That follows from the definition and the elementary fact : if $P \xrightarrow{u} Q \xrightarrow{v} R$ are continuous mappings of locally compact spaces and u is surjective, then $v \circ u$ is proper if and only u and v are so.

The second equality of (ii) is then a consequence of 10.5(ii). To prove the first one, we note that by 10.4(3) the functors $(g \circ f)^!$ and $f^! \circ g^!$ are right adjoints respectively to $R(g \circ f)_!$ and $Rg_! \circ Rf_!$, which have just been shown to be equal.

10.7 PROPOSITION . *Let*

(1)
$$\begin{array}{ccc} X' & \xrightarrow{q'} & X \\ \downarrow{\scriptstyle h} & & \downarrow{\scriptstyle f} \\ Y' & \xrightarrow{q} & Y \end{array}$$

be a cartesian diagram of spaces. Then

(2) $$q^*f_!A = h_!q'^*A \quad \text{in} \quad Sh(Y') \qquad (A \in Sh(X))$$

(3) $$q^*Rf_!A^\bullet = Rh_!q'^*A^\bullet \ \text{in}\ D(Y') \qquad (A^\bullet \in D(X))$$

(4) $$Rh_*q'^!B^\bullet = q^!Rf_*B^\bullet \ \text{in} \ D(Y') \qquad (B^\bullet \in D(X))$$

Proof. Let $V \subset Y$ be open. If $C \subset f^{-1}(V)$ is such that the restriction of f to $C \to V$ is proper, then so is the restriction of h to $q'^{-1}(C) \to q^{-1}(V)$. From this we get natural homomorphisms

(5) $$\alpha : q^*f_!A \to h_!q'^*A$$

(6) $$\beta : q^*Rf_!A^\bullet \to Rh_!q'^*A^\bullet \ .$$

It suffices then to show that α (resp. β) is an isomorphism of the stalks (resp. stalk cohomology groups). Let $y' \in Y'$ and $y = q(y')$. By VI,2.4 we have

(7) $$(q^*f_!A)_{y'} = (f_!A)_y = \Gamma_c(f^{-1}y;A^\bullet)$$

(8) $$(h_!q'^*A)_{y'} = \Gamma_c(h^{-1}y';q^*A)$$

It follows from (7) and (8) that we have

(9) $$H^\bullet(q^*Rf_!A^\bullet)_{y'} = H^\bullet(Rf_!A^\bullet)_y = \mathbf{H}_c^\bullet(f^{-1}y;A^\bullet)$$

(10) $$H^\bullet(Rh_!q^*A^\bullet)_{y'} = \mathbf{H}_c^\bullet(h^{-1}y';q'^*A^\bullet)$$

Since q' induces a homeomorphism of $h^{-1}y'$ onto $f^{-1}y$, this shows that $\alpha_{y'}$ (resp. $\beta_{y'}$) is an isomorphism (resp. quasi-isomorphism). This proves (2) and (3).

Exchanging Y and Y' we also get

$$(11) \qquad f^*Rq_!A^\bullet = Rq'_!h^*A^\bullet , \quad (A^\bullet \in D(Y'))$$

which implies (4) by adjonction.

10.8 PROPOSITION. (i) *Let* $\mathcal{A} \in Sh(X)$ *be c-soft and* $\mathcal{B} \in Sh(Y)$ *be flat. Then*

$$(1) \qquad f_!(\mathcal{A} \otimes f^*\mathcal{B}) = f_!\mathcal{A} \otimes \mathcal{B} , \text{ in } Sh(Y) .$$

(ii) *We have in* D(Y)

$$(2) \qquad Rf_!(A^\bullet \overset{L}{\otimes} f^*\mathcal{B}^\bullet) = Rf_!A^\bullet \overset{L}{\otimes} \mathcal{B}^\bullet \quad (A^\bullet \in D(X), \mathcal{B}^\bullet \in D(Y)) .$$

Proof. (i) There is an obvious map from the right hand side of (1) to the left hand side. Therefore we need only to check that the stalks at $y \in Y$ are naturally isomorphic. We have, by VI,2.6 :

$$(3) \qquad (f_!\mathcal{A} \otimes \mathcal{B})_y = (f_!\mathcal{A})_y \otimes \mathcal{B}_y = \Gamma_c(f^{-1}y;\mathcal{A}) \otimes \mathcal{B}_y ,$$

$$(4) \qquad (f_!(\mathcal{A} \otimes f^*\mathcal{B}))_y = \Gamma_c(f^{-1}y;\mathcal{A} \otimes f^*\mathcal{B})) .$$

But $f^*\mathcal{B}|f^{-1}y$ is the constant sheaf with stalk $\mathcal{B}_y$. The equality of the last terms in (3) and (4) is then obvious if $\mathcal{B}_y$ is a free R-module. By assumption, $\mathcal{B}_y$ is flat hence, by a theorem of D. Lazard, is an inductive limit of free R-modules (see e.g. [3:p.14]). Since cohomology with compact supports commutes with inductive limits [5:II,4.12.1], (i) follows.

(ii) We may assume $A^\bullet$ to be c-soft and $\mathcal{B}^\bullet$ to be flat. Then $A^\bullet \otimes f^*\mathcal{B}^\bullet$ is also c-soft (6.5). We are therefore reduced to proving

$$f_!(A^\bullet \otimes f^*\mathcal{B}^\bullet) = f_!A^\bullet \otimes \mathcal{B}^\bullet ,$$

when $A^\bullet$ is c-soft and $\mathcal{B}^\bullet$ is flat. But this follows from (i).

10.9 *Remarks.* (1) To prove (2) we need only to use (1) when $\mathcal{A}$ is c-soft, in which case (3) and (4) already follow from 7.12.

(2) We note that (1) would be false in general if $f_!$ were replaced by f_* (unless f is proper, of course). In fact, if X is a discrete set and f the projection onto a point, then $\mathcal{B}$ is just an R-module and $f_*(\mathcal{A} \otimes \mathcal{B})$ is the direct product of the modules $\mathcal{A}_x \otimes \mathcal{B}$ $(x \in X)$, while $f_*(\mathcal{A}) \otimes \mathcal{B}$ is $(\Pi_x \mathcal{A}_x) \otimes \mathcal{B}$, and these are different in general if X is infinite.

10.10 PROPOSITION. (i) *Let* $\mathcal{K} \in \mathrm{Sh}(X)$ *be flat and c-soft. Then we have, in* Sh(X)

$$f^!_{\mathcal{K}}(\mathcal{H}om(\mathcal{A},\mathcal{B})) = \mathcal{H}om(f^*\mathcal{A},f^!_{\mathcal{K}}\mathcal{B}) \quad , \quad (\mathcal{A},\mathcal{B} \in \mathrm{Sh}(Y)) \ . \tag{1}$$

(ii) *In* D(X), *we have the equality*

$$f^{\cdot !}R\mathcal{H}om^{\cdot}(\mathcal{A}^{\cdot},\mathcal{B}^{\cdot}) = R\mathcal{H}om^{\cdot}(f^*\mathcal{A}^{\cdot},f^{\cdot !}\mathcal{B}^{\cdot}) \quad , \quad (\mathcal{A}^{\cdot},\mathcal{B}^{\cdot} \in D(Y)) \tag{2}$$

Proof : (i) We have to establish a natural isomorphism between the spaces of sections of the two sides of (1) over an arbitrary open subset U of X . It follows directly from the definitions that we have

$$f^!_{\mathcal{K}}\mathcal{B}|U = (f|U)^!_{\mathcal{K}_U}(\mathcal{B}) \ . \tag{3}$$

As a consequence, it suffices to prove the equality (1) for U = X. By 10.3(3) .

$$f_*\mathcal{H}om(f^*\mathcal{A},f^!_{\mathcal{K}}\mathcal{B}) = \mathcal{H}om(\mathcal{A},f_*f^!_{\mathcal{K}}\mathcal{B}) \ , \tag{4}$$

hence, since 7.14(2) implies $\mathcal{H}om(f_!\mathcal{K},\mathcal{B}) = f_*f^{\cdot !}\mathcal{B}$:

$$f_*\mathcal{H}om(f^*\mathcal{A},f^!_{\mathcal{K}}\mathcal{B}) = \mathcal{H}om(\mathcal{A},\mathcal{H}om(f_!\mathcal{K},\mathcal{B})). \tag{5}$$

Using 10.2 we get

$$f_*\mathcal{H}om(f^*\mathcal{A},f^!_{\mathcal{K}}\mathcal{B}) = \mathcal{H}om(\mathcal{A} \otimes f_!\mathcal{K},\mathcal{B}) \ . \tag{6}$$

On the other hand, from 7.14(2) and 10.3 we also get

(7) $$f_* f^!_K(\mathcal{Hom}(A,B)) = \mathcal{Hom}(f_! K, \mathcal{Hom}(A,B)) = \mathcal{Hom}(f_! K \otimes A, B) ,$$

whence

(8) $$f_* f^!_K(\mathcal{Hom}(A,B)) = f_* \mathcal{Hom}(f^*A, f^!_K B) .$$

The equality of the spaces of sections on Y of the two sides in (8) then gives the required equality.

(ii) We may assume $\mathcal{B}^\bullet$ to be injective and $A^\bullet$ flat. Then $f^!\mathcal{B}^\bullet$ and $\mathcal{Hom}^\bullet(A^\bullet,\mathcal{B}^\bullet)$ are also injective, so that we are reduced to proving

(9) $$f^! \mathcal{Hom}^\bullet(A^\bullet,\mathcal{B}^\bullet) = \mathcal{Hom}^\bullet(f^*A^\bullet, f^!\mathcal{B}^\bullet) ,$$

but this follows from (1) .

10.11 THEOREM . *In* D(X) *and* D(Y) *we have the following equalities*

(1) $$f^! \mathcal{D}^\bullet_Y = \mathcal{D}^\bullet_X .$$

(2) $$f^! \mathcal{D}^\bullet_Y A^\bullet = \mathcal{D}^\bullet_X f^* A^\bullet , \quad \mathcal{D}^\bullet_Y Rf_! \mathcal{B}^\bullet = Rf_* \mathcal{D}^\bullet_X \mathcal{B}^\bullet \qquad (A^\bullet \in D(Y), \mathcal{B}^\bullet \in D(X)) .$$

Proof: (1) Let $g : Y \to pt$ be the projection of Y to a point. By 7.18, we have

(3) $$\mathcal{D}^\bullet_Y = g^! R_{pt} , \quad \mathcal{D}^\bullet_X = (g \circ f)^! R_{pt} .$$

Therefore (1) follows from (3) and the first equality in 10.6(ii). The equality 10.10(2) with $\mathcal{B} = \mathcal{D}^\bullet_Y$, then gives

(4) $$f^! R\mathcal{Hom}^\bullet(A^\bullet, \mathcal{D}^\bullet_Y) = R\mathcal{Hom}^\bullet(f^*A^\bullet, \mathcal{D}^\bullet_X) ,$$

which is the first equality in (2) . The Verdier duality 10.4(2) yields

(5) $$R\mathcal{Hom}^\bullet(Rf_! \mathcal{B}^\bullet, \mathcal{D}^\bullet_Y) = Rf_* R\mathcal{Hom}^\bullet(\mathcal{B}^\bullet, f^! \mathcal{D}^\bullet_Y) .$$

Since $f^! \mathcal{D}_Y^\cdot = \mathcal{D}_X^!$ by (1), this proves the second part of (2) .

B. *Stratified maps*

In this section, except in 10.14, X and Y are equipped with topological (unrestricted) stratifications (2.1) $\mathfrak{X}$ and $\mathfrak{Y}$ and $f : X \to Y$ is a stratified map ([6:1.2], see below).

10.12 The continuous map $f : X \to Y$ is *stratified* if the following conditions are fulfilled:

(i) For every connected component S of a stratum of $\mathfrak{Y}$, the space $f^{-1}S$ is a union of connected components of strata of $\mathfrak{X}$.

(ii) Let $y \in Y$ and S the stratum of $\mathfrak{Y}$ containing y . Then there exists a neighborhood U of y in S , a stratified space F and a stratification preserving homeomorphism $F \times U \xrightarrow{\sim} f^{-1}U$ which transforms the projection to U onto f .

In (ii) it is of course understood that the stratification of $U \times F$ is the product by U of the given stratification $\mathcal{F}$ of F . This condition implies in particular that if T is a connected component of a stratum of $\mathfrak{Y}$, then $f(X) \cap T$ is open in T (possibly empty).

Also, if f is a closed inclusion, then each stratum of $\mathfrak{X}$ is the intersection of X with a stratum of $\mathfrak{Y}$ of the same dimension.

For brevity, we shall say f is *algebraic* if X,Y are complex algebraic varieties, $\mathfrak{X}$ and $\mathfrak{Y}$ are defined by algebraic subsets, and f is a morphism of algebraic varieties.

10.13 We shall say that X, endowed with the topological stratification $\mathfrak{X}$, is compactifiable if it is embeddable as an open dense set in a compact space $\bar{X}$ of the same dimension admitting a topological stratification $\bar{\mathfrak{X}}$ whose trace on X is a refinement of $\mathfrak{X}$. Then $(\bar{X},\bar{\mathfrak{X}})$ is said to be a compactification of $(X,\mathfrak{X})$. If X is already compact, then $(X,\mathfrak{X})$ is a compactification of itself. The main point of this definition however is that if $(X,\mathfrak{X})$ is algebraic, then it is known to be compactifiable.

LEMMA . *Assume* $(X,\mathfrak{X})$ *to be compactifiable. Let* $A^\bullet \in DGS(X)$ *be* $\mathfrak{X}$ *- cc. Then* $\mathbb{H}_c^\bullet(X;A^\bullet)$ and $\mathbb{H}^\bullet(X;A^\bullet)$ *are finitely generated* $(i \in \mathbb{Z})$.

If X is compact, this follows from 3.10 . Let now $(\bar{X},\bar{\mathfrak{X}})$ be a compactification of $(X,\mathfrak{X})$ and $i : X \to \bar{X}$ the inclusion map. Clearly $i_!A^\bullet$, (which is extension by zero), is $\bar{\mathfrak{X}}$ - cc, hence

$$\mathbb{H}_c^\bullet(X;A^\bullet) = \mathbb{H}_c^\bullet(\bar{X};i_!A^\bullet) ,$$

is finitely generated (3.10). By 3.11 , $Ri_*A^\bullet$ is $\mathfrak{X}$ - cc, hence

$$\mathbb{H}^\bullet(X;A^\bullet) = \mathbb{H}^\bullet(X;Ri_*A^\bullet) \qquad (i \in \mathbb{Z})$$

is also finitely generated.

We shall need the following complement to 3.8.

10.14 LEMMA . *Let* $(M,\mathfrak{M})$ *be a stratified pseudomanifold,* Y *a locally contractible space and* π *the projection of* $X = Y \times M$ *onto* Y . *We let* $\mathfrak{X}$ *be the stratification of* X *product of* $\mathfrak{M}$ *by* Y . *Let* $S^\bullet \in DGS(X)$ *be* $\mathfrak{X}$ *- clc. Let* σ *be the projection of* X *onto* M .

(i) *If* Y *is moreover contractible the adjonction morphism* $\sigma^* R\sigma_* S^\bullet \to S^\bullet$ *is a q.i.*

(ii) *The sheaves* $R\pi_*S^\bullet$ *and* $R\pi_!S^\bullet$ *are clc on* Y . *If* $Z \subset Y$ *is open, contractible then the restriction map induces an isomorphism*

$$(1) \qquad \mathbb{H}^\bullet(\pi^{-1}Z;S^\bullet) = \mathbb{H}^\bullet(\pi^{-1}y;S^\bullet) \qquad (y \in Y) .$$

In particular

$$(2) \qquad (R^i\pi_*S^\bullet)_y = \mathbb{H}^i(\pi^{-1}y;S^\bullet) \qquad (y \in Y;\ i \in \mathbb{Z})$$

If $y \in Y$ *and* T *is a compact contractible neighborhood of* y , *then the restriction map induces an isomorphism*

$$(3) \qquad \mathbb{H}_c^\bullet(\pi^{-1}T;S^\bullet) = \mathbb{H}_c^\bullet(\pi_y^{-1};S^\bullet) .$$

Proof . (i) Consider first the case where M is a manifold with the trivial stratification. Let $x = (y,z) \in X$. Then

$$(4)\quad \mathcal{H}^{\bullet}(\sigma^*R\sigma_*\mathcal{S}^{\bullet})_x = \mathcal{H}^{\bullet}(R\sigma_*\mathcal{S}^{\bullet})_z = \varinjlim \mathbb{H}^{\bullet}(U;R\sigma_*\mathcal{S}^{\bullet}) = \varinjlim \mathbb{H}^{\bullet}(U \times Y;\mathcal{S}^{\bullet}) ,$$

where U runs through the neighborhoods of z in M . If U is contractible, then $\mathcal{H}^{\bullet}(\mathcal{S}^{\bullet})$ is constant on $U \times Y$ and since $U \times Y$ is contractible, the hypercohomology spectral sequence and 1.11(b) give

$$(5)\qquad \mathbb{H}^{\bullet}(U \times Y;\mathcal{S}^{\bullet}) = \mathcal{H}^{\bullet}\mathcal{S}^{\bullet}_x .$$

Therefore $\mathcal{H}^{\bullet}(\sigma^*R\sigma_*\mathcal{S}^{\bullet})_x = \mathcal{H}^{\bullet}\mathcal{S}^{\bullet}_x$, as claimed. To prove (i) in general, we use for $\mathfrak{M}$ our usual notation for stratifications and proceed by induction on $k \geqq 1$. We have a commutative diagram

$$(6)\qquad \begin{array}{ccccc} U_k \times Y & \xrightarrow{\ i\ } & U_{k+1} \times Y & \xleftarrow{\ j\ } & S_{n-k} \times Y \\ \downarrow{\scriptstyle \sigma'} & & \downarrow{\scriptstyle \sigma} & & \downarrow{\scriptstyle \sigma''} \\ U_k & \xrightarrow{\ i'\ } & U_{k+1} & \xleftarrow{\ j'\ } & S_{n-k} \end{array}$$

where i and j stand for i_k and j_k . We assume that $R\sigma_*\mathcal{S}^{\bullet}_k \to \mathcal{S}^{\bullet}_k$ is an isomorphism and want to prove that the same is true with k replaced by k+1 . Adjunction gives a natural morphism of distinguished triangles

$$(7)\qquad \begin{array}{ccc} \sigma^*R\sigma_*j_!j^!\mathcal{S}^{\bullet}_{k+1} & \longrightarrow & \sigma^*R\sigma_*\mathcal{S}^{\bullet}_{k+1} \\ {\scriptstyle [1]}\nwarrow & & \swarrow \\ & \sigma^*R\sigma_*Ri_*\mathcal{S}^{\bullet}_k & \end{array} \quad \longrightarrow \quad \begin{array}{ccc} j_!j^!\mathcal{S}^{\bullet}_{k+1} & \longrightarrow & \mathcal{S}^{\bullet}_{k+1} \\ {\scriptstyle [1]}\nwarrow & & \swarrow \\ & Ri_*\mathcal{S}^{\bullet}_k & \end{array}$$

therefore it suffices to check that

$$(8)\qquad \sigma^*R\sigma_*Ri_*\mathcal{S}^{\bullet}_k \longrightarrow Ri_*\mathcal{S}^{\bullet}_k \quad \text{and} \quad \sigma^*R\sigma_*j_!j^!\mathcal{S}^{\bullet}_{k+1} \longrightarrow j_!j^!\mathcal{S}^{\bullet}_{k+1}$$

are isomorphisms. By 3.13 and the induction assumption, we have

$$(9)\quad \sigma^*R\sigma_*Ri_*\mathcal{S}^{\bullet}_k = \sigma^*Ri'_*R\sigma'_*\mathcal{S}^{\bullet}_k = Ri_*\sigma'^*R\sigma'_*\mathcal{S}^{\bullet}_k = Ri_*\mathcal{S}^{\bullet}_k$$

which gives the first part of (8). The induction hypothesis implies that $R\sigma^{!}_{*}S^{\bullet}_{k}$ is $\mathfrak{M}$ - clc. Hence so is $Ri^{!}_{*}R\sigma^{!}_{*}S^{\bullet}_{k}$ by 3.9 . By (9), $Ri_{*}S^{\bullet}_{k+1}$ is therefore $\mathfrak{X}$ - clc. Using the right hand side triangle of (7), in which $S^{\bullet}_{k+1}$ is also $\mathfrak{X}$ - clc, we find then that $j^{!}S^{\bullet}_{k+1}$ is clc. Since the stratum S_{n-k} of M is a manifold, we have

$$\sigma^{*}R\sigma_{*}j_{!}j^{!}S^{\bullet}_{k+1} = \sigma^{*}j^{!}_{*}R\sigma''_{*}j^{!}S^{\bullet}_{k+1} = j_{*}\sigma''^{*}R\sigma''_{*}j^{!}S^{\bullet}_{k+1} = j_{!}j^{!}S^{\bullet}_{k+1}$$

which proves the second part of (8) .

(ii) Let $j_{y} : \pi^{-1}y \to \pi^{-1}Z$ be the inclusion. Then (i) implies

$$(10) \qquad R\sigma_{*}S^{\bullet} = j^{*}_{y}S^{\bullet} \ , \quad S^{\bullet} = \sigma^{*}j^{*}_{y}S^{\bullet} \ ,$$

hence (1) follows from the Vietoris-Begle theorem (3.13, Remark). Since $(R^{i}\pi_{*}S^{\bullet})_{y}$ is the inductive limit of the $\mathbb{H}^{i}(\pi^{-1}Z;S^{\bullet})$, where Z runs through a fundamental set of neighborhoods of y, this also proves (2) and shows that $R\pi_{*}S^{\bullet}$ is clc. If T is compact contractible, then (3) follows from (10) and the usual Vietoris-Begle theorem. By 10.5, we may view $R\pi_{!}S^{\bullet}$ as the Leray sheaf $R\pi_{c}S^{\bullet}$ of π . Again $H^{\bullet}(R\pi_{c}S^{\bullet})_{y}$ is the inductive limit of the $\mathbb{H}^{\bullet}_{c}(\pi^{-1}T;S^{\bullet})$, and it follows that $R\pi_{!}S^{\bullet}$ is clc.

Remark. In this lemma "Y locally contractible" means : every $y \in Y$ has a fundamental system of open neighborhoods and of closed neighborhoods which are contractible.

10.15 LEMMA . *Assume* f *is a closed inclusion and let* $B^{\bullet} \in$ DGS(Y) *be* $\mathfrak{Y}$ *- clc (resp.* $\mathfrak{Y}$ *- cc). Then* $f^{!}B^{\bullet}$ *is* $\mathfrak{X}$ *- clc (resp.* $\mathfrak{X}$ *- cc).*

Proof. Let $A^{\bullet} = f^{!}B^{\bullet}$. We use the usual notation (2.1) for $\mathfrak{X}$ and prove by induction on $k \geqq 1$ that $A^{\bullet}_{k}$ is $\mathfrak{X}$ - clc (resp. $\mathfrak{X}$ - cc). Since U_{1} is a union of connected components of strata of Y of the same dimension as X , our assertion for $A^{\bullet}_{1}$ follows from 3.10.
Assume it is proved for some $k \geqq 1$. We consider the distinguished triangle

$$\begin{array}{ccc} j_{k!}j^{!}_{k}A^{\bullet}_{k+1} & \longrightarrow & A^{\bullet}_{k+1} \\ & {\scriptstyle [1]}\nwarrow \quad \swarrow & \\ & Ri_{k*}A^{\bullet}_{k} & \end{array}$$

Since $A^\bullet_k$ is $\mathfrak{X}$ - clc (resp. $\mathfrak{X}$ - cc) on U_k , the same is true for $Ri_{k*}A^\bullet_k$ on U_{k+1} by 3.9, 3.11 . There remains then to see that $j^!_k A^\bullet$ is clc (resp. with finitely generated stalk cohomology) on a connected component Z of S_{n-k} . Let $\ell : Z \to Y$ be the inclusion. Then $\ell = f \circ j_k$, hence $\ell^! = j^!_k \circ f^!$ (1.9). We have then $j^!_k A^\bullet_{k+1} = \ell^! B^\bullet$, and the latter has the required properties by 3.10, 3.11 , in view of the fact that Z is a connected component of a stratum of $\mathfrak{Y}$.

10.16 THEOREM . *Let* $A^\bullet \in DGS(X)$ *and* $B^\bullet \in DGS(Y)$.

(i) *If* $B^\bullet$ *is* $\mathfrak{Y}$ *- clc (resp.* $\mathfrak{Y}$ *- cc), then* $f^*B^\bullet$ *is* $\mathfrak{X}$ *- clc (resp.* $\mathfrak{X}$ *- cc).*

(ii) *If* $A^\bullet$ *is* $\mathfrak{X}$ *- clc, then* $Rf_*A^\bullet$ *and* $Rf_!A^\bullet$ *are* $\mathfrak{Y}$ *- clc.*

(iii) *If* $A^\bullet$ *is* $\mathfrak{X}$ *- clc and* $\mathbb{H}^i_c(f^{-1}y;A^\bullet)$ *is finitely generated for every* $y \in Y$ *and* $i \in \mathbb{Z}$*, then* $Rf_!A^\bullet$ *is* $\mathfrak{Y}$ *- cc.*

(iv) *For* $0 \leq k \leq m = \dim Y$*, let* $T_k = Y_k - Y_{k-1}$*,* ℓ_k *the inclusion of* $f^{-1}(T_{m-k})$ *in* X*, and* h_k *the restriction of* f *to* $f^{-1}(T_{m-k})$*. Assume that* $A^\bullet$ *is* $\mathfrak{X}$ *- clc and that for every* k *(* $0 \leq k \leq m$*),* $y \in T_{m-k}$ *and* $i \in \mathbb{Z}$ *,* $\mathbb{H}^i(f^{-1}y;\ell^!_k A^\bullet)$ *is finitely generated. Then* $Rf_*A^\bullet$ *is* $\mathfrak{Y}$ *- cc.*

(v) *If* $A^\bullet$ *is* $\mathfrak{X}$ *- cc and every fibre of* f *is compactifiable* (10.13), *then* $Rf_!A^\bullet$ *and* $Rf_*A^\bullet$ *are* $\mathfrak{Y}$ *- cc.*

Proof: (i) Let S be a stratum of $\mathfrak{Y}$. Then $B^\bullet$ is clc (resp. with finitely generated stalk cohomology in each degree) on S . The same is therefore true for $f^*B^\bullet$ on $f^{-1}S$, hence also a fortiori for the restriction of $f^*B^\bullet$ to any component of a stratum of $\mathfrak{X}$ contained in $f^{-1}S$. Since any connected component of a stratum of X is caught in this way, this proves (i).

(ii), (iii), (iv). It follows from 7.12 (or VI,2.6) that

$$(1) \qquad H^\bullet(Rf_!A^\bullet)_y = \mathbb{H}^\bullet_c(f^{-1}y;A^\bullet) \qquad (y \in Y) .$$

Let $y \in Y$ and let U,F be as in 10.12(ii). We assume that U is a ball of dimension d . Let h be the restriction of f to $f^{-1}(U)$. By 10.7(3), $(Rf_!A^\bullet)_{|U} = Rh_!(A^\bullet_{|f^{-1}U})$, and 10.14(ii) shows that $Rh_!(A^\bullet_{|f^{-1}U})$ is clc on U. As a consequence it is $\mathfrak{Y}$ - clc. In view of (1), it is then $\mathfrak{Y}$ - cc under the assumptions of (iii).

We now prove, by induction on $k \geqq 1$, that $\mathcal{B}^\bullet = Rf_*\mathcal{A}^\bullet$ is $\mathfrak{Y}$ - clc on $V_k = Y - Y_{m-k}$, with finitely generated stalk cohomology under the assumptions of (iv). Let $y \in V_1$. We use the notation of 10.12(ii). Then U is also a neighborhood of y in Y. Since $f^{-1}V_1$ is open in X, we have by 10.14(ii)

$$(2) \qquad (R^i f_*\mathcal{A}^\bullet)_y = \mathbb{H}^i(f^{-1}y;\mathcal{A}^\bullet) = \mathbb{H}^i(f^{-1}y;\ell_1^!\mathcal{A}^\bullet) \quad (i \in \mathbb{Z})$$

and $Rf_*\mathcal{A}^\bullet$ is clc on U. This proves (ii) and (iv) for $(Rf_*\mathcal{A}^\bullet)_1$. Assume they hold for some $k \geqq 1$. As usual, we consider the distinguished triangle

$$\begin{array}{ccc} j_{k!}j_k^!\mathcal{B}^\bullet_{k+1} & \longrightarrow & \mathcal{B}^\bullet_{k+1} \\ & {\scriptstyle [1]}\nwarrow \quad \swarrow & \\ & Ri_{k*}\mathcal{B}^\bullet_k & \end{array}$$

By assumption $\mathcal{B}^\bullet$ is $\mathfrak{Y}$ - clc (resp. $\mathfrak{Y}$ - cc in case (iv) on V_k. Then the same is true for $Ri_{k*}\mathcal{B}^\bullet_k$ by 3.9 ,3.11 . It suffices therefore to prove that $j_k^!\mathcal{B}^\bullet_{k+1}$ is clc on T_{m-k} (resp. with finitely generated stalk cohomology in case (iv)). By 10.7(4), we have on T_{m-k}

$$(3) \qquad j_k^!\mathcal{B}^\bullet = j_k^! Rf_*\mathcal{A}^\bullet = Rh_*\ell^!\mathcal{A}^\bullet .$$

By 10.15, $\ell^!\mathcal{A}^\bullet$ is $\mathfrak{X}$ - clc. Since T_{m-k} is a manifold, we know by the case $k = 1$ that $Rh_*\ell^!\mathcal{A}^\bullet$ satisfies our conditions.

(v) By 10.13 the assumption of (iii) are fulfilled, and so are those of (iv) by 10.15 and 10.13 .

10.17 THEOREM (i) *The functors* f^* *and* $f^!$ *map* $D^b(\mathfrak{Y})$ *into* $D^b(\mathfrak{X})$ *and satisfy the relations*

$$(1) \qquad f^! = \mathbb{D}^\bullet_X f^* \mathbb{D}^\bullet_Y , \quad f^* = \mathbb{D}^\bullet_X f^! \mathbb{D}^\bullet_Y , \quad \mathbb{D}^\bullet_X f^* = f^! \mathbb{D}^\bullet_Y , \quad \mathbb{D}^\bullet_X f^! = f^* \mathbb{D}^\bullet_Y$$

(ii) *Assume* f *to be proper or algebraic* (10.12), *or more generally that every fibre of* f *is compactifiable* (10.13). *Then* Rf_* *and* $Rf_!$ *map*

$D^b(\mathfrak{X})$ *into* $D^b(\mathfrak{Y})$ *and satisfy the relations*

$$(2) \qquad Rf_! = \mathcal{D}^\bullet_Y Rf_* \mathcal{D}^\bullet_X \ , \ Rf_* = \mathcal{D}^\bullet_Y Rf_! \mathcal{D}^\bullet_X \ , \ \mathcal{D}^\bullet_Y Rf_* = Rf_! \mathcal{D}^\bullet_X \ , \ \mathcal{D}^\bullet_Y Rf_! = Rf_* \mathcal{D}^\bullet_X \ .$$

Proof. (i) We already know by 10.16 that f^* maps $D^b(\mathfrak{Y})$ into $D^b(\mathfrak{X})$ and by 10.11 that the third relation of (1) is satisfied. Using the biduality in Y we get then the first one. Since $\mathcal{D}^\bullet_X$ and $\mathcal{D}^\bullet_Y$ preserve $D^b(\mathfrak{X})$ and $D^b(\mathfrak{Y})$ resp. (8.7), this shows that $f^!$ maps $D^b(\mathfrak{Y})$ into $D^b(\mathfrak{X})$. The other two relations in (1) follow from the others and biduality.

(ii) By 10.16, Rf_* and $Rf_!$ map $D^b(\mathfrak{X})$ into $D^b(\mathfrak{Y})$. By 10.11, the last relation of (2) holds. The other ones then follow by biduality.

We conclude this section with an elementary fact which is a companion to 8.6:

10.18 PROPOSITION . *Let* $A^\bullet, B^\bullet \in DGS^b(X)$. *If they are* $\mathfrak{X}$ *- clc (resp.* $\mathfrak{X}$ *- cc) then so is* $A^\bullet \overset{L}{\otimes} B^\bullet$.

Proof. Let $x \in X$ and U an open neighborhood of x in the stratum of X containing it. We have to prove that $A^\bullet \overset{L}{\otimes} B^\bullet$ is clc (resp. clc with finitely generated stalk cohomology) on U. We may assume U to be a ball. $A^\bullet$ and $B^\bullet$ are then cohomologically constant on U. By 8.1 we may assume they are complexes of constant sheaves. We may then replace $B^\bullet$ by a bounded flat left resolution by constant sheaves. We have then

$$A^\bullet(U) \overset{L}{\otimes} B^\bullet(U) = A^\bullet(U) \otimes B^\bullet(U) = A^\bullet_y \otimes B^\bullet_y = A^\bullet_y \overset{L}{\otimes} B^\bullet_y \ , \quad (y \in U)$$

From this it follows that $A^\bullet \otimes B^\bullet$ is a constant DGS on U. This shows first that $A^\bullet \overset{L}{\otimes} B^\bullet$ is $\mathfrak{X}$ - clc. Moreover

$$H^\bullet(A^\bullet \overset{L}{\otimes} B^\bullet)_y = H^\bullet(A^\bullet_y \otimes B^\bullet_y)$$

is finitely generated if $H^\bullet(A^\bullet_y)$ and $H^\bullet(B^\bullet_y)$ are finitely generated (as follows from the Tor-spectral sequence), and the proposition follows.

C. *Some identities on products*

The spaces X *and* Y *are as in part* A *in* 10.19 to 10.21 *and as in part* B *from* 10.22 *on. We let* $p : X \times Y \to X$ *and* $q : X \times Y \to Y$ *denote the canonical projections.*

10.19 THEOREM . *Let* $\mathcal{A}^\bullet \in DGS^b(X)$, $\mathcal{B}^\bullet \in DGS^b(Y)$. *Then*

(1) $$R\Gamma_c(X;\mathcal{A}^\bullet) \overset{L}{\otimes} R\Gamma_c(Y;\mathcal{B}^\bullet) = R\Gamma_c(X \times Y;\ p^*\mathcal{A}^\bullet \overset{L}{\otimes} q^*\mathcal{B}^\bullet) .$$

Proof. We consider the cartesian diagram

(2) $$\begin{array}{ccc} X \times Y & \xrightarrow{\ q\ } & Y \\ \Big\downarrow p & & \Big\downarrow s \\ X & \xrightarrow{\ r\ } & pt \end{array}$$

By 10.8, we have

(3) $$Rq_!(p^*\mathcal{A}^\bullet \overset{L}{\otimes} q^*\mathcal{B}^\bullet) = Rq_!p^*\mathcal{A}^\bullet \overset{L}{\otimes} \mathcal{B}^\bullet .$$

From 10.7(3) we get

(4) $$Rq_!p^*\mathcal{A}^\bullet = s^*Rr_!\mathcal{A}^\bullet .$$

Using 10.8 again, we derive

$$Rs_!(s^*Rr_!\mathcal{A}^\bullet \overset{L}{\otimes} \mathcal{B}^\bullet) = Rr_!\mathcal{A}^\bullet \overset{L}{\otimes} Rs_!\mathcal{B}^\bullet$$

which, combined with (3) and 10.6, gives

(5) $$Rr_!\mathcal{A}^\bullet \overset{L}{\otimes} Rs_!\mathcal{B}^\bullet = R(s \circ q)_!(p^*\mathcal{A}^\bullet \overset{L}{\otimes} q^*\mathcal{B}^\bullet)$$

But this is just another way to write (1) .

10.20 *Remarks* (a) 10.19 is a generalization of the usual Künneth rule to compute cohomology with compact supports of a product. To see this assume that $A^\bullet = R_X$ and $B^\bullet = R_Y$. Then $p^*A^\bullet \overset{L}{\otimes} q^*B^\bullet = R_{X\times Y}$ and (5) gives

$$(1) \qquad R\Gamma_c(R_X) \overset{L}{\otimes} R\Gamma_c(R_Y) = R\Gamma_c(R_{X\times Y}) ,$$

which provides the Künneth relation.

(b) Let Z be a third space and, changing our notation slightly, consider instead of (2) the cartesian diagram

$$\begin{array}{ccc} X\times_Z Y & \xrightarrow{\ q\ } & Y \\ \downarrow{\scriptstyle p} & & \downarrow{\scriptstyle s} \\ X & \xrightarrow{\ r\ } & Z \end{array}$$

Then (5) is again valid, with the same proof. For this formula, see also [10: Exp. 3, Prop 4.5] or [4: V.3].

(c) Assume that $x \in X$ and $y \in Y$ have some neighborhoods U and V respectively such that

$$(2) \qquad \mathbb{H}^\bullet_c(U;A^\bullet) = H^\bullet f^!_x A^\bullet , \ \mathbb{H}^\bullet_c(V;B^\bullet) = H^\bullet f^!_y B^\bullet .$$

$$(3) \qquad \mathbb{H}^\bullet_c(U\times V; p^*A^\bullet \overset{L}{\otimes} q^*B^\bullet) = H^\bullet f^!_{(x,y)}(p^*A^\bullet \overset{L}{\otimes} q^*B^\bullet) .$$

Then 10.19(1) yields in this case

$$(4) \qquad f^!_x A^\bullet \overset{L}{\otimes} f^!_y B^\bullet = f^!_{(x,y)}(p^*A^\bullet \overset{L}{\otimes} q^*B^\bullet) \quad (x \in X,\ y \in Y) .$$

Note that this assumption is fulfilled when X and Y are endowed with unrestricted stratifications $\mathfrak{X},\mathfrak{Y}$ such that $A^\bullet$ and $B^\bullet$ are $\mathfrak{X}$-clc and $\mathfrak{Y}$-clc respectively (3.10) .

10.21 PROPOSITION. *Assume that* Y *is locally contractible. Then*

$$(1) \qquad p^*R\mathcal{H}om^\bullet(A^\bullet,B^\bullet) = R\mathcal{H}om^\bullet(p^*A^\bullet,p^*B^\bullet) \quad (A^\bullet,B^\bullet \in D^b(X)) .$$

This assertion is local. We may therefore assume Y to be contractible and locally contractible. Note that if X is stratified, the proposition is easily proved since 10.3 and 10.14(i) give

$$R\mathcal{H}om^{\bullet}(p^*\mathcal{A}^{\bullet},p^*\mathcal{B}^{\bullet}) = p^*Rp_*R\mathcal{H}om^{\bullet}(p^*\mathcal{A}^{\bullet},p^*\mathcal{B}^{\bullet}) = p^*R\mathrm{Hom}^{\bullet}(\mathcal{A}^{\bullet},Rp_*(p^*\mathcal{B}^{\bullet}))$$

and we have $Rp_*(p^*\mathcal{B}^{\bullet}) = \mathcal{B}^{\bullet}$ by 3.13(15).

In the general case, the following argument is due to N. Spaltenstein.

It follows from 3.13(4) that in $Sh(X\times Y)$

$$(2) \qquad \mathcal{H}om(p^*\mathcal{A},p^*\mathcal{B}) = p^*\mathcal{H}om(\mathcal{A},\mathcal{B}) \qquad (\mathcal{A},\mathcal{B} \in Sh(X)).$$

Let now $\mathcal{B} \in Sh(X)$ be injective. If $U \subset X$, $V \subset Y$ are open, with V contractible, the Vietoris-Begle theorem (3.13, Remark) implies in particular that $p^*\mathcal{B}$ is acyclic for $\Gamma(U\times V;\quad)$. It follows then easily that if $\mathcal{A} = \oplus_{i\in I}R_{U_i}$ for some family $(U_i)_{i\in I}$ of open subsets of X, then $p^*\mathcal{B}$ is also acyclic for $\mathcal{H}om(p^*\mathcal{A};\quad) = \Pi_{i\in I}\mathcal{H}om(R_{U_i\times Y};\quad)$. Thus (1) reduces to (2) when $\mathcal{A}^{\bullet}$ is a single degree complex which is a direct sum of sheaves of the form R_U, U open in X. In particular (1) holds in this case. As every sheaf on X is a quotient of such a sheaf, the general case of (1) follows then from [7:1.7.1] (or rather its counterpart for contravariant functors).

10.22 LEMMA. (i) *Let* $\mathcal{A}^{\bullet} \in DGS^b(X)$, $\mathcal{B}^{\bullet} \in DGS^b(Y)$, *and let the index* k *correspond to the filtration on* $X\times Y$ *induced by* $\mathfrak{X}$. *Assume that* $\mathcal{A}^{\bullet}$ *is* $\mathfrak{X}$ - *clc*. *Then*

$$(1) \qquad Ri_{k*}(p^*\mathcal{A}^{\bullet})_k \overset{L}{\otimes} (q^*\mathcal{B}^{\bullet})_{k+1} = Ri_{k*}((p^*\mathcal{A}^{\bullet})_k \overset{L}{\otimes} (q^*\mathcal{B}^{\bullet})_k).$$

(ii) *Let* $\mathcal{A}^{\bullet},\mathcal{B}^{\bullet} \in DGS(X)$, *where* $\mathcal{A}^{\bullet}$ *is* $\mathfrak{X}$ - *clc and* $\mathcal{B}^{\bullet}$ *is clc. Then*

$$(2) \qquad (Ri_{k*}\mathcal{A}^{\bullet}_k) \overset{L}{\otimes} \mathcal{B}^{\bullet}_{k+1} = Ri_{k*}(\mathcal{A}^{\bullet}_k \overset{L}{\otimes} \mathcal{B}^{\bullet}_k) .$$

Proof. (c) We shall use the following easy consequence of the Vietoris-Begle theorem (3.13,Remark). If Z is a topological space, $\mathcal{S}^{\bullet} \in DGS^b(Z)$ and

$$(3)\qquad \begin{array}{ccc} Z\times(0,1) & & \\ & \searrow^{\phi} & \\ i\downarrow & & Z \\ & \nearrow_{\psi} & \\ Z\times[0,1) & & \end{array}$$

are the obvious maps, then

$$(4)\qquad Ri_*\phi^*\mathcal{S}^\bullet \;=\; \psi^*\mathcal{S}^\bullet \;.$$

It is clear that (1) holds over $U_k \times Y$. Let now $x \in S_{n-k}$, and $U = B \times \overset{\circ}{c}(L)$ a distinguished neighborhood of x in X . We prove that (1) holds over $U \times Y$. We have a commutative diagram

$$(5)\qquad \begin{array}{ccc} B\times \overset{\circ}{c}(L)^* \times Y & \xrightarrow{\;i_k\;} & B\times \overset{\circ}{c}(L)\times Y \\ \Big\| & & \uparrow \pi \\ B\times(0,1)\times L\times Y & \xrightarrow{\;i\;} & B\times[0,1)\times L\times Y \end{array}$$

The map π is proper, hence $\pi_* = \pi_!$.
By 10.8 we get

$$(6)\qquad Ri_{k*}(p^*\mathcal{A}^\bullet)_k \overset{L}{\otimes} (q^*\mathcal{B}^\bullet)_{k+1} = R\pi_*(Ri_*(p^*\mathcal{A}^\bullet)_k) \overset{L}{\otimes} (q^*\mathcal{B}^\bullet)_{k+1} =$$
$$= R\pi_*(Ri_*(p^*\mathcal{A}^\bullet)_k \overset{L}{\otimes} \pi^*(q^*\mathcal{B}^\bullet)_{k+1}) \;.$$

On the other hand,

$$(7)\qquad Ri_{k*}((p^*\mathcal{A}^\bullet)_k \overset{L}{\otimes} (q^*\mathcal{B}^\bullet)_k) = Ri_{k*}((p^*\mathcal{A}^\bullet)_k \overset{L}{\otimes} i_k^*((q^*\mathcal{B}^\bullet)_{k+1})) =$$
$$= R\pi_*(Ri_*((p^*\mathcal{A}^\bullet)_k \overset{L}{\otimes} i^*\pi^*(q^*\mathcal{B}^\bullet)_{k+1})) \;.$$

Thus we need only to prove:

$$(8)\qquad Ri_*(p^*\mathcal{A}^\bullet)_k \overset{L}{\otimes} \pi^*(q^*\mathcal{B}^\bullet)_{k+1} = Ri_*((p^*\mathcal{A}^\bullet)_k \overset{L}{\otimes} i^*\pi^*(q^*\mathcal{B}^\bullet)_{k+1}) .$$

Let $Z = B \times L \times Y$, and let ϕ , ψ be the projections defined by (3). Let

$$S^\bullet = R\phi_*(p^*A^\bullet)_k \ , \quad T^\bullet = R\psi_*(\pi^*(q^*B^\bullet)_{k+1}) \ .$$

By 10.14(i) we have

$$(p^*A^\bullet)_k = \phi^*S^\bullet \ , \quad \pi^*(q^*B^\bullet)_{k+1} = \psi^*T^\bullet \ .$$

From (4), we then get

$$Ri_*(p^*A^\bullet)_k = Ri_*(\phi^*S^\bullet) = \psi^*S^\bullet \ .$$

We have therefore

(9) $$Ri_*(p^*A^\bullet)_k \overset{L}{\otimes} \pi^*(q^*B^\bullet)_{k+1} = \psi^*S^\bullet \overset{L}{\otimes} \psi^*T^\bullet \ .$$

On the other hand

$$(p^*A^\bullet)_k \overset{L}{\otimes} i^*\pi^*(q^*B^\bullet)_{k+1} = \phi^*S^\bullet \overset{L}{\otimes} i^*\psi^*T^\bullet = \phi^*S^\bullet \overset{L}{\otimes} \phi^*T^\bullet = \phi^*(S^\bullet \overset{L}{\otimes} T^\bullet)$$

and, by (4)

(10) $$Ri_*(\phi^*(S^\bullet \overset{L}{\otimes} T^\bullet)) = \psi^*(S^\bullet \overset{L}{\otimes} T^\bullet) \ .$$

Since the right hand sides of (9) and (10) are equal, this proves (i).

(ii) We may take $A^\bullet$ flabby and $B^\bullet$ flat. Then

$$Ri_{k*}A_k^\bullet \overset{L}{\otimes} B_{k+1}^\bullet = i_{k*}A_k^\bullet \otimes B_{k+1}^\bullet \ , \ Ri_{k*}(A_k^\bullet \overset{L}{\otimes} B_k^\bullet) = Ri_{k*}(A_k^\bullet \otimes B_k^\bullet).$$

There are natural morphisms

(11) $$i_{k*}A_k^\bullet \otimes B_{k+1}^\bullet \to i_{k*}(A_k^\bullet \otimes B_k^\bullet) \to Ri_{k*}(A_k^\bullet \otimes B_k^\bullet) \ .$$

and this gives a natural morphism

(12) $$Ri_{k*}A_k^{\bullet} \overset{L}{\otimes} \mathcal{B}_{k+1}^{\bullet} \to Ri_{k*}(A_k^{\bullet} \overset{L}{\otimes} \mathcal{B}_k^{\bullet}) .$$

It remains to check that, under the assumptions of (ii), (12) is an isomorphism in D(X). This is local, hence we may assume that X is contractible. By 8.1, we may then assume that $\mathcal{B}^{\bullet}$ is a complex of constant sheaves. In this case (ii) is a special case of (i), with Y = pt .

10.23 PROPOSITION . *Let* $A^{\bullet},\mathcal{B}^{\bullet},\mathcal{C}^{\bullet} \in DGS^b(X)$. *Assume that* $A^{\bullet}$ *is* $\mathfrak{X}$ - *cc,* $\mathcal{B}^{\bullet}$ *is* $\mathfrak{X}$ - *clc and* $\mathcal{C}^{\bullet}$ *is clc. Then*

(1) $$R\mathcal{H}om^{\bullet}(A^{\bullet},\mathcal{B}^{\bullet}) \overset{L}{\otimes} \mathcal{C}^{\bullet} = R\mathcal{H}om^{\bullet}(A^{\bullet},\mathcal{B}^{\bullet} \overset{L}{\otimes} \mathcal{C}^{\bullet}) .$$

Proof. We note first that, for any $A^{\bullet},\mathcal{B}^{\bullet},\mathcal{C}^{\bullet} \in D^b(X)$, there is a natural morphism

(2) $$\nu : R\mathcal{H}om^{\bullet}(A^{\bullet},\mathcal{B}^{\bullet}) \overset{L}{\otimes} \mathcal{C}^{\bullet} \to R\mathcal{H}om^{\bullet}(A^{\bullet},\mathcal{B}^{\bullet} \overset{L}{\otimes} \mathcal{C}^{\bullet}) .$$

In fact we may assume $\mathcal{C}^{\bullet}$ to be flat and $\mathcal{B}^{\bullet}$ to be injective. We then define ν as the composite map

(3) $$\mathcal{H}om^{\bullet}(A^{\bullet},\mathcal{B}^{\bullet}) \otimes \mathcal{C}^{\bullet} \to \mathcal{H}om^{\bullet}(A^{\bullet},\mathcal{B}^{\bullet} \otimes \mathcal{C}^{\bullet}) \to R\mathcal{H}om^{\bullet}(A^{\bullet},\mathcal{B}^{\bullet} \otimes \mathcal{C}^{\bullet}) .$$

We shall prove that (1) is implemented by ν .

Our statement is local on X. We may therefore assume X to be contractible. Then $\mathcal{H}^{\bullet}\mathcal{C}^{\bullet}$ is constant. By 8.1, $\mathcal{C}^{\bullet}$ may be replaced by a bounded complex of constant sheaves, hence also by a bounded complex of constant flat sheaves. We assume this from now on and have to prove that ν induces an isomorphism

(4) $$R\mathcal{H}om^{\bullet}(A^{\bullet},\mathcal{B}^{\bullet}) \otimes \mathcal{C}^{\bullet} = R\mathcal{H}om^{\bullet}(A^{\bullet},\mathcal{B}^{\bullet} \otimes \mathcal{C}^{\bullet}) .$$

We check first that ν is an isomorphism in the case where $A^{\bullet}$ *is* clc *with finitely generated stalk cohomology and* $\mathcal{B}^{\bullet},\mathcal{C}^{\bullet}$ *are arbitrary*. As the problem is local, we may assume that X is contractible. Then $\mathcal{H}^{\bullet}A^{\bullet}$ is constant, and by 8.1 we may replace $A^{\bullet}$ by a complex of constant sheaves with finitely generated free stalks [3: §3,no 5, Prop.7]. This resolution can be used to compute $R\mathcal{H}om^{\bullet}$ [5: II,7.4]. It remains then only to notice that for $A,\mathcal{B},\mathcal{C} \in Sh(X)$ we clearly have

(4) $$Hom(A,B) \otimes C = Hom(A,B \otimes C)$$

if A is a finite direct sum of copies of R_X .

In the general case we argue as usual by induction on k, in the notation of 2.1 . Consider the relation

$(5)_k$ $$RHom^{\bullet}(A_k^{\bullet},B_k^{\bullet}) \otimes C_k^{\bullet} = RHom^{\bullet}(A_k^{\bullet},B_k^{\bullet} \otimes C_k^{\bullet}) .$$

The special case discussed above implies in particular that $(5)_1$ holds. Assume it is true for some $k \geqq 1$. We want to prove $(5)_{k+1}$. Set $i = i_k$ and $j = j_k$. From the distinguished triangle

(6) $$\begin{array}{ccc} j_!j^!B_{k+1}^{\bullet} & \longrightarrow & B_{k+1}^{\bullet} \\ & {\scriptstyle [1]}\nwarrow \quad \swarrow & \\ & Ri_*B_k^{\bullet} & \end{array}$$

we get by applying ν a morphism of the distinguished triangle

(7) $$\begin{array}{ccc} RHom^{\bullet}(A_{k+1}^{\bullet},j_!j^!B_{k+1}^{\bullet}) \overset{L}{\otimes} C_{k+1}^{\bullet} & \longrightarrow & RHom^{\bullet}(A_{k+1}^{\bullet},B_{k+1}^{\bullet}) \overset{L}{\otimes} C_{k+1}^{\bullet} \\ & {\scriptstyle [1]}\nwarrow \quad \swarrow & \\ & RHom^{\bullet}(A_{k+1}^{\bullet},Ri_*B_k^{\bullet}) \overset{L}{\otimes} C_{k+1}^{\bullet} & \end{array}$$

into the distinguished triangle

(8) $$\begin{array}{ccc} RHom^{\bullet}(A_{k+1}^{\bullet},j_!j^!(B_{k+1}^{\bullet}) \overset{L}{\otimes} C_{k+1}^{\bullet}) & \longrightarrow & RHom^{\bullet}(A_{k+1}^{\bullet},B_{k+1}^{\bullet} \overset{L}{\otimes} C_{k+1}^{\bullet}) \\ & {\scriptstyle [1]}\nwarrow \quad \swarrow & \\ & RHom^{\bullet}(A_{k+1}^{\bullet},Ri_*(B_k^{\bullet}) \overset{L}{\otimes} C_{k+1}^{\bullet} . & \end{array}$$

It suffices to show that it is an isomorphism on the left edge. Using successively 10.3(1), 10.22(ii), $(5)_k$, 10.3(1)and 10.22(ii) again,we get

$$(9)\qquad RHom^{\bullet}(A^{\bullet}_{k+1},Ri_{*}B^{\bullet}_{k}) \overset{L}{\otimes} C^{\bullet}_{k+1} = Ri_{*}RHom^{\bullet}(A^{\bullet}_{k},B^{\bullet}_{k}) \overset{L}{\otimes} C^{\bullet}_{k+1} =$$

$$= Ri_{*}(RHom^{\bullet}(A^{\bullet}_{k},B^{\bullet}_{k}) \overset{L}{\otimes} C^{\bullet}_{k}) = Ri_{*}RHom^{\bullet}(A^{\bullet}_{k},B^{\bullet}_{k} \overset{L}{\otimes} C^{\bullet}_{k}) =$$

$$= RHom^{\bullet}(A^{\bullet}_{k+1},Ri_{*}(B^{\bullet}_{k} \overset{L}{\otimes} C^{\bullet}_{k})) = RHom^{\bullet}(A^{\bullet}_{k+1},Ri_{*}(B^{\bullet}_{k}) \overset{L}{\otimes} C^{\bullet}_{k+1}) .$$

This proves the equality of the two bottom vertices of the triangles (7) and (8). For the upper left vertices of (7) and (8), we remark first that $Rj_{*} = j_{*} = j_{!} = Rj_{!}$ since S_{n-k} is closed in U_{k+1}. Note also that by assumption $j^{*}A^{\bullet}_{k+1}$ is clc with finitely generated stalk cohomology. Using sucessively 10.3(1), 10.8(ii), the special case discussed above, 10.3(1) and 10.8(ii) again, we get

$$(10)\qquad RHom^{\bullet}(A^{\bullet}_{k+1},j_{!}j^{!}B^{\bullet}_{k+1}) \overset{L}{\otimes} C^{\bullet}_{k+1} = Rj_{!}RHom^{\bullet}(j^{*}A^{\bullet}_{k+1},j^{!}B^{\bullet}_{k+1}) \overset{L}{\otimes} C^{\bullet}_{k+1} =$$

$$= Rj_{!}(RHom^{\bullet}(j^{*}A^{\bullet}_{k+1},j^{!}B^{\bullet}_{k+1}) \overset{L}{\otimes} j^{*}C^{\bullet}_{k+1}) = Rj_{!}RHom^{\bullet}(j^{*}A^{\bullet}_{k+1},j^{!}B^{\bullet}_{k+1} \overset{L}{\otimes} j^{*}C^{\bullet}_{k+1})$$

$$= RHom^{\bullet}(A^{\bullet}_{k+1},j_{!}(j^{!}B^{\bullet}_{k+1} \overset{L}{\otimes} j^{*}C^{\bullet}_{k+1}) = RHom^{\bullet}(A^{\bullet}_{k+1},j_{!}j^{!}B^{\bullet}_{k+1} \overset{L}{\otimes} C^{\bullet}_{k+1}).$$

Thus the upper left vertices of (7) and (8) are also isomorphic, and it follows that $(5)_{k+1}$ holds.

10.24 Our next goal is to prove 10.25, but before doing that, we need to show the existence of a natural morphism α from the left hand side to the right hand side of 10.25(1).

(a) We first consider the case where $A^{\bullet} = R_X$. We want therefore a morphism

$$(1)\qquad \alpha : p^{*}D^{\bullet}_{X} \overset{L}{\otimes} q^{*}B^{\bullet} \to q^{!}B^{\bullet} .$$

We may assume $B^{\bullet}$ to be flat. Then so is $q^{*}B^{\bullet}$ and we can erase the L on the left hand side. Fix a flat c-soft resolution $K^{\bullet}$ of R_X and an injective resolution $I^{\bullet}$ of R . The complex $L^{\bullet} = p^{*}K^{\bullet}$ is a flat resolution of $R_{X \times Y}$ whose restriction to any fibre of q is c-soft . It can therefore be used to compute $q^{!}$ (VI,§3). On the other hand, $I^{\bullet}_{Y} = s^{*}I^{\bullet}$, where s is the projection of Y to a point, is a resolution of R_Y and, since $\mathcal{B}^\cdot$ is flat, $\mathcal{B}^\cdot \overset{L}{\otimes} I_Y^\cdot = \mathcal{B}^\cdot \otimes I_Y^\cdot = \mathcal{B}^\cdot$. Therefore, if $\mathcal{J}^\cdot$ is an injective resolution of $\mathcal{B}^\cdot \otimes I_Y^\cdot$, then $q_{\mathcal{L}^\cdot}^!(\mathcal{J}^\cdot)$ represents $q^!\mathcal{B}^\cdot$. In particular there is a natural morphism $q_{\mathcal{L}^\cdot}^!(\mathcal{B}^\cdot \otimes I_Y^\cdot) \to q_{\mathcal{L}^\cdot}^!(\mathcal{J}^\cdot) = q^!\mathcal{B}^\cdot$. To show the existence of α it suffices therefore to prove the existence of a natural morphism

$$(2) \qquad p^*\mathcal{D}_X^\cdot \otimes q^*\mathcal{B}^\cdot \to q_{\mathcal{L}^\cdot}^!(\mathcal{B}^\cdot \otimes I_Y^\cdot) .$$

For this, it is sufficient to define, for $U \subset X$ and $V \subset Y$ open, a natural map

$$(3) \qquad p^*\mathcal{D}_X^\cdot(U \times V) \times q^*\mathcal{B}^\cdot(U \times V) \to q_{\mathcal{L}^\cdot}^!(\mathcal{B}^\cdot \otimes I_Y^\cdot)(U \times V) .$$

In view of the definition of a morphism of sheaves, we need actually only to define a natural map

$$(4) \quad \mathcal{D}_X^\cdot(U) \times \mathcal{B}^\cdot(V) \to q_{\mathcal{L}^\cdot}^!(\mathcal{B}^\cdot \otimes I_Y^\cdot)(U \times V) = \mathrm{Hom}_{Sh(Y)}(q_!(\mathcal{L}_{U \times V}^\cdot), \mathcal{B}^\cdot \otimes I_Y^\cdot).$$

In order to do so, it is sufficient to give, for every connected open subset V' of V, a natural map

$$(5) \qquad \mathcal{D}_X^\cdot(U) \times \mathcal{B}^\cdot(V) \to \mathrm{Hom}_R(q_!(\mathcal{L}_{U \times V}^\cdot)(V'), (\mathcal{B}^\cdot \otimes I_Y^\cdot)(V')),$$

or equivalently

$$(6) \qquad \mathcal{D}_X^\cdot(U) \times \mathcal{B}^\cdot(V) \times q_!(\mathcal{L}_{U \times V}^\cdot)(V') \to (\mathcal{B}^\cdot \otimes I_Y^\cdot)(V') .$$

Since $\mathcal{L}^\cdot = q^*\mathcal{K}^\cdot$ and V' is connected, we have

$$(7) \quad q_!(\mathcal{L}_{U \times V}^\cdot(V')) = \Gamma_{\Phi_{V'}}(X \times V'; (p^*\mathcal{K}^\cdot)_{U \times V}) = \Gamma_{\Phi_{V'}}(X \times V'; p^*(\mathcal{K}_U^\cdot)) =$$
$$= \Gamma_c(X; \mathcal{K}_U^\cdot) .$$

Therefore (6) is equivalent to a map

$$(8) \qquad \mathcal{D}_X^\cdot(U) \times \mathcal{B}^\cdot(V) \times \Gamma_c(X; \mathcal{K}_U^\cdot) \to (\mathcal{B}^\cdot \otimes I_Y^\cdot)(V') .$$

But $\mathcal{D}^\bullet_X(U) = \mathrm{Hom}(\Gamma_c(X;\mathcal{K}^\bullet_U),I^\bullet)$. Therefore we can define the required map in (8) as the composite of the map

(9) $$\mathcal{D}^\bullet_X(U) \times \mathcal{B}^\bullet(V) \times \Gamma_c(X;\mathcal{K}^\bullet_U) \to \mathcal{B}^\bullet(V') \otimes I^\bullet$$

given by

(10) $$(f,b,\gamma) \longmapsto (b|_{V'}) \otimes f(\gamma) \quad (f \in \mathcal{D}^\bullet_X(U),\ b \in \mathcal{B}^\bullet(V),\ \gamma \in \Gamma_c(X;\mathcal{K}^\bullet_U)),$$

with the natural map from $\mathcal{B}^\bullet(V') \otimes I^\bullet$ to $(\mathcal{B}^\bullet \otimes I^\bullet_Y)(V')$.

(b) We can now define a morphism

$$\mu : p^*\mathbf{D}^\bullet_X A^\bullet \otimes q^*\mathcal{B}^\bullet \to R\mathcal{H}om^\bullet(p^*A^\bullet,q^!\mathcal{B}^\bullet),\quad (A^\bullet \in D^b(X),\ \mathcal{B}^\bullet \in D^b(Y)) .$$

We have, by 7.8 and 10.21, quasi-isomorphisms

(11) $$\beta : p^*\mathbf{D}^\bullet_X A^\bullet \overset{L}{\otimes} q^*\mathcal{B}^\bullet = p^*R\mathcal{H}om^\bullet(A^\bullet,\mathcal{D}^\bullet_X) \overset{L}{\otimes} q^*\mathcal{B}^\bullet = R\mathcal{H}om^\bullet(p^*A^\bullet,p^*\mathcal{D}^\bullet_X) \overset{L}{\otimes} q^*\mathcal{B}^\bullet,$$

and from 10.23(2), a natural morphism

(12) $$\nu : R\mathcal{H}om^\bullet(p^*A^\bullet,p^*\mathcal{D}^\bullet_X) \overset{L}{\otimes} q^*\mathcal{B}^\bullet \to R\mathcal{H}om^\bullet(p^*A^\bullet,p^*\mathcal{D}^\bullet_X \overset{L}{\otimes} q^*\mathcal{B}^\bullet) .$$

We then define α as the composition of $\nu \circ \beta$ with the morphism

$$R\mathcal{H}om^\bullet(p^*A^\bullet,p^*\mathcal{D}^\bullet_X \overset{L}{\otimes} q^*\mathcal{B}^\bullet) \to R\mathcal{H}om^\bullet(p^*A^\bullet,q^!\mathcal{B}^\bullet)$$

induced by the map α defined in (a) .

10.25 THEOREM . *Let* $A^\bullet \in \mathrm{DGS}^b(X)$ *be* $\mathfrak{X}$ *- clc and* $\mathcal{B}^\bullet \in \mathrm{DGS}^b(Y)$ *be* $\mathfrak{Y}$ *- cc . Then*

(1) $$p^*\mathbf{D}^\bullet_X A^\bullet \overset{L}{\otimes} q^*\mathcal{B}^\bullet = R\mathcal{H}om^\bullet(p^*A^\bullet,q^!\mathcal{B}^\bullet) .$$

Proof. In view of 10.24, we need only to check (1) on the stalks. Let $x \in X$, $y \in Y$. We have

(2) $$(p^*\mathbf{D}^\bullet_X A^\bullet \overset{L}{\otimes} q^*\mathcal{B}^\bullet)_{(x,y)} = (\mathbf{D}^\bullet_X A^\bullet)_x \overset{L}{\otimes} \mathcal{B}^\bullet_y .$$

It follows easily from 3.10 that

(3) $$(\mathbb{D}_X^{\bullet}A^{\bullet})_x = \mathrm{RHom}^{\bullet}(f_x^{!}A^{\bullet},R) \; .$$

Thus for the left hand side of (1) we get

(4) $$(p^*\mathbb{D}_X^{\bullet}A^{\bullet} \overset{L}{\otimes} q^*\mathcal{B}^{\bullet})_{(x,y)} = \mathrm{RHom}^{\bullet}(f_x^{!}A^{\bullet},R) \overset{L}{\otimes} \mathcal{B}_y^{\bullet} \; .$$

Consider now the right hand side of (1). Since $\mathcal{B}^{\bullet}$ is $\mathfrak{Y}$ - cc , 10.17(i) gives

(5) $$q^{!}\mathcal{B}^{\bullet} = \mathbb{D}_{X\times Y}^{\bullet}q^*\mathbb{D}_Y^{\bullet}\mathcal{B}^{\bullet} = R\mathcal{H}om^{\bullet}(q^*\mathbb{D}_Y^{\bullet}\mathcal{B}^{\bullet},\mathcal{D}_{X\times Y}^{\bullet})$$

By 10.2, we have therefore

(6) $$R\mathcal{H}om^{\bullet}(p^*A^{\bullet},q^{!}\mathcal{B}^{\bullet}) = R\mathcal{H}om^{\bullet}(p^*A^{\bullet},R\mathcal{H}om^{\bullet}(q^*\mathbb{D}_Y^{\bullet}\mathcal{B}^{\bullet},\mathcal{D}_{X\times Y}^{\bullet})) =$$

$$= R\mathcal{H}om^{\bullet}(p^*A^{\bullet} \overset{L}{\otimes} q^*\mathbb{D}_Y^{\bullet}\mathcal{B}^{\bullet},\mathcal{D}_{X\times Y}^{\bullet}) = \mathbb{D}_{X\times Y}^{\bullet}(p^*A^{\bullet} \overset{L}{\otimes} q^*\mathbb{D}_Y^{\bullet}\mathcal{B}^{\bullet}) \; .$$

As in (3), we derive from 3.10 :

(7) $$(\mathbb{D}_{X\times Y}^{\bullet}(p^*A^{\bullet} \overset{L}{\otimes} q^*\mathbb{D}_Y^{\bullet}\mathcal{B}^{\bullet}))_{(x,y)} = \mathrm{RHom}^{\bullet}(f_{(x,y)}^{!}(p^*A^{\bullet} \overset{L}{\otimes} q^*\mathbb{D}_Y^{\bullet}\mathcal{B}^{\bullet}),R) \; .$$

Using then 10.20(4),10.2 and 8.11, we get

(8) $$R\mathcal{H}om^{\bullet}(p^*A^{\bullet},q^{!}\mathcal{B}^{\bullet})_{(x,y)} = \mathrm{RHom}^{\bullet}(f_{(x,y)}^{!}(p^*A^{\bullet} \overset{L}{\otimes} q^*\mathbb{D}_Y^{\bullet}\mathcal{B}^{\bullet}),R) =$$

$$= \mathrm{RHom}^{\bullet}(f_x^{!}A^{\bullet} \overset{L}{\otimes} f_y^{!}\mathbb{D}_Y^{\bullet}\mathcal{B}^{\bullet},R) =$$

$$= \mathrm{RHom}^{\bullet}(f_x^{!}A^{\bullet},\mathrm{RHom}^{\bullet}(f_y^{!}\mathbb{D}_Y^{\bullet}\mathcal{B}^{\bullet},R)) =$$

$$= \mathrm{RHom}^{\bullet}(f_x^{!}A^{\bullet},\mathcal{B}_y^{\bullet}) \; .$$

Thus we need only to check that

(9) $$\mathrm{RHom}^{\bullet}(f_x^{!}A^{\bullet},R) \overset{L}{\otimes} \mathcal{B}_y^{\bullet} = \mathrm{RHom}^{\bullet}(f_x^{!}A^{\bullet},\mathcal{B}_y^{\bullet}) \; .$$

This is a statement about complexes of modules. By [3:§3,no 5, Prop.7] the complex $\mathcal{B}^{\bullet}_Y$ is q.i. to a complex $N^{\bullet}$ of finitely generated free modules. Then $I^{\bullet} \otimes N^{\bullet}$ is an injective resolution of $N^{\bullet}$, and (9) is equivalent to

$$\text{(10)} \qquad \mathrm{Hom}^{\bullet}(f_X^{!}\mathcal{A}^{\bullet}, I^{\bullet}) \otimes N^{\bullet} = \mathrm{Hom}^{\bullet}(f_X^{!}\mathcal{A}^{\bullet}, I^{\bullet} \otimes N^{\bullet}),$$

which obviously holds since $N^{\bullet}$ consists of finitely generated free modules.

10.26 COROLLARY .*We have* $p^*\mathcal{D}^{\bullet}_X \overset{L}{\otimes} q^*\mathcal{D}^{\bullet}_Y = \mathcal{D}^{\bullet}_{X\times Y}$.

Proof. In 10.25(1) take $\mathcal{A}^{\bullet} = R_X$ and $\mathcal{B}^{\bullet} = \mathcal{D}^{\bullet}_Y$. Since $q^{!}\mathcal{D}^{\bullet}_Y = \mathcal{D}^{\bullet}_{X\times Y}$, (10.11) we get

$$p^*\mathcal{D}^{\bullet}_X \overset{L}{\otimes} q^*\mathcal{D}^{\bullet}_Y = R\mathcal{H}om^{\bullet}(p^*R^{\bullet}_X, \mathcal{D}^{\bullet}_{X\times Y}) = R\mathcal{H}om^{\bullet}(R_{X\times Y}, \mathcal{D}^{\bullet}_{X\times Y}) = \mathcal{D}^{\bullet}_{X\times Y} .$$

10.27 COROLLARY . *Let* $Y = X$ *and* $d : X \to X \times X$ *be the diagonal map. Then*

$$\text{(1)} \qquad d^{!}(p^*\mathcal{D}^{\bullet}_X\mathcal{A}^{\bullet} \overset{L}{\otimes} q^*\mathcal{B}^{\bullet}) = R\mathcal{H}om^{\bullet}(\mathcal{A}^{\bullet}, \mathcal{B}^{\bullet}), \quad (\mathcal{A}^{\bullet}, \mathcal{B}^{\bullet} \in DGS^b(X)).$$

Proof. Since $p \circ d = q \circ d = \mathrm{Id.}$, 10.6 and 10.10(ii) show that

$$\text{(2)} \qquad d^{!}R\mathcal{H}om^{\bullet}(p^*\mathcal{A}^{\bullet}, p^{!}\mathcal{B}^{\bullet}) = R\mathcal{H}om^{\bullet}(\mathcal{A}^{\bullet}, \mathcal{B}^{\bullet}) .$$

The corollary then follows by applying $d^{!}$ to both sides of 10.25(1).

Remark. The statement of 10.25 is borrowed from [10:Exp.9,p.44], where it is called "Théorème du noyau" and is proved under somewhat more general assumptions. My original argument was an induction on strata as in 10.22 or in many places in these Notes. The simpler proof given here was communicated to me by N. Spaltenstein.

REFERENCES

[1] A. BOREL, *The Poincaré duality in generalized manifolds*, Mich. Math. J. **4** (1957) 227-239

[2] A. BOREL and J.C. MOORE, *Homology theory for locally compact spaces*, Mich. Math. J. **7** (1960) 137-159

[3] N. BOURBAKI, Algèbre, Chap. 10 : Algèbre homologique, Masson, Paris 1980

[4] G. BREDON, Sheaf Theory, McGraw Hill 1967

[5] R. GODEMENT, Topologie algébrique et Théorie des faisceaux, Act. Sci. Ind. **1252**, Hermann, Paris 1958

[6] M. GORESKY and R. MACPHERSON, *Intersection homology* II, Inv. Math. **71** (1983) 77-129

[7] R. HARTSHORNE, Residues and duality, Lecture Notes in Math. **20**, Springer-Verlag 1966

[8] B. IVERSEN, Cohomology of sheaves, Notes, Aarhus University, 1984

[9] L. KAUP and M.S. KEANE, *Induktive Limiten endlich erzeugter freier Moduln*, Manuscripta Math. **1** (1969) 9-21

[10] Séminaire Heidelberg-Strasbourg 1966-67, IRMA

[11] J.L. VERDIER, *Dualité dans la cohomologie des espaces localement compacts*, Séminaire Bourbaki, Exp. **300** (1965-66)

[12] J.L. VERDIER, *Catégories dérivées*, Etat 0, SGA 4 1/2, Lecture Notes in Math. **569** (1977), 262-311

[13] J.L. VERDIER, *Classe d'homologie associée à un cycle*, Astérisque **36-37** (1976) 101-151

Intersection Cohomology
Progress in Mathematics, Vol. 50
© 1984 Birkhäuser Boston, Inc.

VI : LES FONCTEURS DE LA CATEGORIE DES FAISCEAUX ASSOCIES A UNE APPLICATION CONTINUE

par Pierre-P. Grivel

Introduction

Lorsque X et Y sont des espaces localement compacts et $f : X \to Y$ est une application continue, Verdier (voir [V]) définit deux foncteurs de la catégorie des faisceaux de A-modules, notés $f_!$ et $f^!$, qui permettent de déterminer une dualité de Poincaré pour les espaces localement compacts. L'actualité de ces foncteurs, connus et utilisés depuis longtemps en géométrie algébrique (voir [H]), provient du fait qu'ils interviennent dans la théorie de la cohomologie d'intersection (voir [GMP]).

Le foncteur $f_!$, foncteur image directe à supports propres, généralise le foncteur $\Gamma_c(X;-)$ des sections à supports compacts et a des propriétés plus régulières que le foncteur image directe f_*. Le foncteur $f^!$ est dans un certain sens un adjoint du foncteur $f_!$.

Dans cette note, émanation d'une partie d'un séminaire tenu à Genève durant le semestre d'été 1983, on se propose de donner une construction détaillée et autant que possible élémentaire de ces deux foncteurs.

Je remercie Nathan Habegger avec qui j'ai eu plusieurs discussions au sujet de ce travail et dont les suggestions m'ont permis d'améliorer certaines démonstrations.

1. LES FONCTEURS f_* et f^*

1.1 Si X est un espace topologique nous désignerons par Ouv(X) l'ensemble des ouverts de X et nous noterons Sh(X) la catégorie abélienne des faisceaux de A-modules sur X, où A est un anneau commutatif unitaire.

Soit $\mathcal{F}$ un faisceau sur X; si $U \in \mathrm{Ouv}(X)$ nous noterons $\Gamma(U;\mathcal{F})$ (resp. $\Gamma_\Phi(U;\mathcal{F})$) le A-module des sections de $\mathcal{F}$ sur U (resp. à supports dans la famille Φ) et si $V \in \mathrm{Ouv}(X)$ tel que $U \subset V$ nous noterons $\rho_{U,V} : \Gamma(V;\mathcal{F}) \to \Gamma(U;\mathcal{F})$ le morphisme de restriction.

1.2 Considérons une application continue $f : X \to Y$. Si $\mathcal{A}$ est un faisceau sur X on vérifie facilement qu'on obtient un faisceau sur Y, noté $f_*\mathcal{A}$, en posant :

$$\Gamma(V;f_*\mathcal{A}) = \Gamma(f^{-1}(V);\mathcal{A}) \quad \text{pour tout } V \in \mathrm{Ouv}(Y)$$

$$\rho_{V,W} = \rho_{f^{-1}(V);f^{-1}(W)} \quad \text{pour tout } V,W \in \mathrm{Ouv}(Y)$$

tels que $V \subset W$.

La correspondance $\mathcal{A} \to f_*\mathcal{A}$ définit évidemment un foncteur $f_* : \mathrm{Sh}(X) \to \mathrm{Sh}(Y)$.

Ce foncteur f_* est exact gauche car le foncteur $\Gamma(U;-)$ est exact gauche pour tout $U \in \mathrm{Ouv}(X)$.

1.3 *Exemples :*

1.3.1 Si Y est un point alors $f_* = \Gamma(X;-)$

1.3.2 Si $j : X \to Y$ est l'inclusion d'un fermé X dans Y alors $j_*\mathcal{A} = \mathcal{A}^Y$ (où $\mathcal{A}^Y$ est l'extension de $\mathcal{A}$ par zéro à Y). Il en résulte que le foncteur j_* est exact.

En effet pour tout $V \in \mathrm{Ouv}(Y)$ considérons l'inclusion $\psi(V) : \Gamma_{\Psi_V}(V \cap X;\mathcal{A}) \to \Gamma(V \cap X;\mathcal{A})$ où Ψ_V est la famille des ensembles C fermés dans $V \cap X$ qui sont fermés dans V. Mais $\psi(V)$ est aussi surjective car si C est fermé dans $V \cap X$ il en résulte que C est fermé dans V puisque X est un sous-espace fermé; donc $C \in \Psi_V$. Ainsi la famille $\{\psi(V)\}_{V \in \mathrm{Ouv}(Y)}$, évidemment compatible avec les restrictions, définit un isomorphisme de faisceaux $\psi : \mathcal{A}^Y \to f_*\mathcal{A}$.

Par contre si $j : X \to Y$ est l'inclusion d'un ouvert X dans Y alors en général $j_*\mathcal{A} \neq \mathcal{A}^Y$. Prenons par exemple l'inclusion j de $X =]0;1[$ dans $Y = [0;1[$ et soit $\mathcal{A} = \mathbb{Z}_X$ le faisceau constant sur X avec fibre $\mathbb{Z}$. Si on calcule la fibre de $j_*\mathcal{A}$ et de $\mathcal{A}^Y$ au point y=0 de Y on obtient $(j_*\mathcal{A})_0 = \mathbb{Z}$ tandis que $(\mathcal{A}^Y)_0 = \{0\}$.

1.4 THEOREME : *Le foncteur $f_* : Sh(X) \to Sh(Y)$ admet un foncteur adjoint à gauche $f^* : Sh(Y) \to Sh(X)$.*

Démonstration : Soit $\mathcal{B}$ un faisceau sur Y et soit $\pi : L\mathcal{B} \to Y$ l'espace étalé sur Y associé à $\mathcal{B}$. Alors par définition $f^*\mathcal{B}$ est le faisceau associé à l'espace étalé $X \times_Y L\mathcal{B} \to X$ obtenu en prenant le produit fibré de π par f.

Il résulte de cette construction que pour tout $x \in X$ on a un isomorphisme entre les fibres

$$(f^*\mathcal{B})_x = \mathcal{B}_{f(x)} \tag{1.4.1}$$

On peut alors montrer ([B],chap. I, §4) que pour tout faisceau $\mathcal{A}$ sur X et tout faisceau $\mathcal{B}$ sur Y on a un isomorphisme

$$\Phi : \mathrm{Hom}_{Sh(X)}(f^*\mathcal{B};\mathcal{A}) \to \mathrm{Hom}_{Sh(Y)}(\mathcal{B};f_*\mathcal{A}) \tag{1.4.2}$$

On en déduit les flèches d'adjonction

$$\begin{aligned} \alpha &: f^*f_* \to 1_{Sh(X)} \\ \beta &: 1_{Sh(Y)} \to f_*f^* \end{aligned} \tag{1.4.3}$$

1.5 COROLLAIRE : *Le foncteur f^* est exact.*

Démonstration : Cela résulte de (1.4.1).

1.6 *Exemple :* Si $j : X \to Y$ est l'inclusion d'un sous-espace X dans Y et si $\mathcal{A}$ est un faisceau sur Y alors $j^*\mathcal{A} = \mathcal{A}_{|X}$ est la restriction de $\mathcal{A}$ à X.

1.7 Soit Inj(X) la sous-catégorie pleine de Sh(X) dont les objets sont les faisceaux injectifs. On notera que la catégorie Inj(X) n'est pas abélienne.

COROLLAIRE : *Le foncteur* f_* *:* $Sh(X) \to Sh(Y)$ *se restreint à un foncteur* f_* *:* $Inj(X) \to Inj(Y)$.

Démonstration : Soit $0 \to \mathcal{A} \to \mathcal{B} \to \mathcal{C} \to 0$ une suite exacte de faisceaux sur Y et soit $\mathcal{I}$ un faisceau injectif sur X. En appliquant successivement à cette suite les foncteurs exacts f^* et $Hom_{Sh(X)}(-;\mathcal{I})$, puis en utilisant l'isomorphisme Φ de (1.4.2) on obtient une suite exacte

$$0 \to Hom_{Sh(Y)}(\mathcal{C};f_*\mathcal{I}) \to Hom_{Sh(Y)}(\mathcal{B};f_*\mathcal{I}) \to Hom_{Sh(Y)}(\mathcal{A};f_*\mathcal{I}) \to 0.$$

Il en résulte que le faisceau $f_*\mathcal{I}$ est injectif.

2. LE FONCTEUR $f_!$

2.1 Soit $f : X \to Y$ une application continue entre espaces localement compacts.

Si V est un ouvert de Y désignons par Φ_V la famille des ensembles C qui sont fermés dans $f^{-1}(V)$ et tels que l'application $f_{|C} : C \to V$ soit propre.

LEMME : *a) Pour tout* $V \in Ouv(Y)$, Φ_V *est une famille de supports sur* $f^{-1}(V)$.

b) Si V et W sont des ouverts de Y tels que $V \subset W$, *l'inclusion* $j : V \to W$ *induit une application* $\tilde{j} : \Phi_W \to \Phi_V$ *donnée par* $\tilde{j}(C) = C \cap f^{-1}(V)$.

Démonstration : a) Soit K un ensemble compact contenu dans V. Si $C,D \in \Phi_V$ alors $C \cup D$ est fermé dans $f^{-1}(V)$ et on a $f^{-1}_{|C\cup D}(K) = f^{-1}_{|C}(K) \subset f^{-1}_{|D}(K)$; il est donc clair que $C \cup D \in \Phi_V$. Si maintenant $C \in \Phi_V$ et D est un ensemble fermé dans $f^{-1}(V)$ tel que $D \subset C$ on a $f^{-1}_{|D}(K) \subset f^{-1}_{|C}(K)$ et puisque $f^{-1}_{|D}(K)$ est fermé il est clair que $D \in \Phi_V$.

b) Si $C \in \Phi_W$ alors $C \cap f^{-1}(V)$ est fermé dans $f^{-1}(V)$; de plus si K est un ensemble compact contenu dans V, donc dans W, on a $f^{-1}_{|C\cap f^{-1}(V)}(K) = f^{-1}_{|C}(K)$ et cet ensemble est compact; donc $C\cap f^{-1}(V) \in \Phi_V$.

<u>2.2</u> Si $\mathcal{A}$ est un préfaisceau sur X il résulte du lemme précédent qu'on définit un préfaisceau sur Y, noté $f_! \mathcal{A}$, en posant

$$\Gamma(V\ f_!\mathcal{A}) = \Gamma_{\Phi_V}(f^{-1}(V);\mathcal{A}) \quad \text{pour tout } V \in \mathrm{Ouv}(Y)$$

$$\rho_{V,W} = \rho_{f^{-1}(V);f^{-1}(W)} \quad \text{pour tout } V,W \in \mathrm{Ouv}(Y)$$

tels que $V \subset W$.

PROPOSITION : *Si $\mathcal{A}$ est un faisceau sur X alors $f_!\mathcal{A}$ est un faisceau sur Y.*

Démonstration : Soit $\{V_i\}_{i\in I}$ une famille d'ouverts de Y ; posons $V = \bigcup_{i\in I} V_i$ et $V_{ij} = V_i \cap V_j$. Alors $\{f^{-1}(V_i)\}_{i\in I}$ est une famille d'ouverts de X telle que $f^{-1}(V) = \bigcup_{i\in I} f^{-1}(V_i)$ et $f^{-1}(V_{ij}) = f^{-1}(V_i) \cap f^{-1}(V_j)$. Puisque $\mathcal{A}$ est un faisceau sur X on a une suite exacte

(2.2.1)

$$0 \rightarrow \Gamma(f^{-1}(V);\mathcal{A}) \overset{\varphi}{\rightarrow} \prod_{i\in I} \Gamma(f^{-1}(V_i);\mathcal{A}) \overset{\psi}{\rightarrow} \prod_{(i,j)\in I\times I} \Gamma(f^{-1}(V_{ij});\mathcal{A})$$

où $\varphi(s) = \left(\rho_{f^{-1}(V_i),f^{-1}(V)}(s)\right)_{i\in I}$ et

$$\psi((s_i)_{i\in I}) = \left(\rho_{f^{-1}(V_{ij}),f^{-1}(V_i)}(s_i) - \rho_{f^{-1}(V_{ij}),f^{-1}(V_j)}(s_j)\right)_{(i,j)\in I\times I}$$

Il suffit donc de vérifier que la suite induite

$$0 \to \Gamma_{\Phi_V}(f^{-1}(V);\mathcal{A}) \xrightarrow{\overline{\varphi}} \prod_{i\in I} \Gamma_{\Phi_{V_i}}(f^{-1}(V_i);\mathcal{A}) \xrightarrow{\overline{\psi}} \prod_{(i,j)\in I\times I} \Gamma_{\Phi_{V_{ij}}}(f^{-1}(V_{ij});\mathcal{A})$$

est encore exacte.

Le seul point non trivial est de voir que $\mathrm{Ker}\,\overline{\psi} \subset \mathrm{Im}\,\overline{\varphi}$. Soit $(s_i)_{i\in I} \in \mathrm{Ker}\,\overline{\psi} \subset \mathrm{Ker}\,\psi$; en vertu de l'exactitude de la suite (2.2.1) il existe $s \in \Gamma(f^{-1}(V);\mathcal{A})$ tel que $\varphi(s) = (s_i)_{i\in I}$. Il faut donc vérifier que le support $|s|$ de s appartient à Φ_V, c'est-à-dire que l'application $f_{||s|} : |s| \to V$ est propre.

Soit un ensemble compact $K \subset V = \bigcup_{i\in I} V_i$. Comme Y est localement compact un petit argument de topologie générale montre qu'on peut trouver un sous-ensemble fini $J \subset I$ et une famille de compacts $\{K_j\}_{j\in J}$ tels que $K_j \subset V_j$ pour tout $j \in J$ et $K \subset \bigcup_{j\in J} K_j$.
Comme $|s_j| = |s| \cap f^{-1}(V_j)$ on a $f^{-1}_{||s|}(K) \subset \bigcup_{j\in J} f^{-1}_{||s_j|}(K_j)$ et ce dernier ensemble est compact car $|s_j| \in \Phi_{V_j}$. Donc $f^{-1}_{||s|}(K)$ est compact.

2.3 Soit $\varphi : \mathcal{A} \to \mathcal{B}$ un morphisme de faisceaux sur X; si $V \in \mathrm{Ouv}(Y)$ et si $s \in \Gamma_{\Phi_V}(f^{-1}(V);\mathcal{A})$ alors $\varphi(f^{-1}(V))(s) \in \Gamma_{\Phi_V}(f^{-1}(V);\mathcal{B})$ car $|\varphi(f^{-1}(V)(s)| \subset |s|$. Il en résulte que la correspondance $\mathcal{A} \to f_!\mathcal{A}$ définit un foncteur $f_! : Sh(X) \to Sh(Y)$. Ce foncteur $f_!$ est exact gauche car le foncteur $\Gamma_{\Phi_V}(f^{-1}(V);-)$ est exact gauche pour tout $V \in \mathrm{Ouv}(Y)$.

2.4 PROPOSITION : *Il existe un morphisme canonique de foncteurs*

$$0 \to f_! \to f_*$$

De plus si l'application $f : X \to Y$ *est propre, les foncteurs* $f_!$ *et* f_* *sont isomorphes.*

Démonstration : Le morphisme de foncteurs est défini par la famille d'inclusions $\{\Gamma_{\Phi_V}(f^{-1}(V);\mathcal{A}) \to \Gamma(f^{-1}(V);\mathcal{A})\}_{V\in \mathrm{Ouv}(Y)}$.
Si f est une application propre, il faut voir que $C \in \Phi_V$ si et seulement si C est un ensemble fermé dans $f^{-1}(V)$.
Soit C un ensemble fermé dans $f^{-1}(V)$ et soit un compact $K \subset V$; alors $f^{-1}(K)$ est un compact contenu dans $f^{-1}(V)$, donc $f^{-1}_{|C}(K) = C \cap f^{-1}(K)$ est compact ; ainsi $C \in \Phi_V$.

2.5 *Exemples :*

2.5.1 Si Y est un point alors $f_! = \Gamma_c(X;-)$ (où c est la famille des ensembles compacts de X; comme X est localement compact, c est une famille paracompactifiante de supports).

2.5.2 Si $j : X \to Y$ est l'inclusion d'un ouvert ou d'un fermé X dans Y alors $j_!A = A^Y$ et le foncteur $f_!$ est exact. (Il est intéressant de rapprocher ce résultat de celui de 1.3.2).

En effet si X est fermé dans Y alors j est une application propre, donc $j_! = j_*$ et le résultat découle de 1.3.2.
Si X est ouvert dans Y, il faut montrer que pour tout $V \in \mathrm{Ouv}(Y)$ on a $\Phi_V = \Psi_V$.

Si $C \in \Phi_V$ alors pour tout compact K contenu dans V, l'ensemble $C \cap K$ est compact dans le sous-espace localement compact V, donc C est fermé dans V. Inversément si $C \in \Psi_V$ et si K est un compact contenu dans V alors $C \cap K$ est compact.

2.6 On peut caractériser les fibres du faisceau $f_!A$ de la façon suivante.

PROPOSITION : *Soit A un faisceau sur X et un point $y \in Y$. Il existe un isomorphisme*

$$\varphi : (f_!A)_y \to \Gamma_c(f^{-1}(y);A_{|f^{-1}(y)})$$

Démonstration : Soit $s_y \in (f_!A)_y$; soit V un voisinage ouvert de y et $s \in \Gamma_{\Phi_V}(f^{-1}(V);A)$ un représentant de s_y. On a $s_{|f^{-1}(y)} \in \Gamma_c\left(f^{-1}(y);A_{|f^{-1}(y)}\right)$; en effet $|s_{|f^{-1}(y)}| = |s| \cap f^{-1}(y)$ est compact puisque $|s| \in \Phi_V$. Comme la valeur de $s_{|f^{-1}(y)}$ est indépendante du choix du représentant de s_y on définit le morphisme φ en posant $\varphi(s_y) = s_{|f^{-1}(y)}$. Montrons que φ est injectif. Soit $s_y \in (f_!A)_y$ représenté par $s \in \Gamma_{\Phi_V}(f^{-1}(V);A)$. Supposons que $\varphi(s_y) = 0$. Alors

$s_{|f^{-1}(y)} = 0$, donc $|s| \cap f^{-1}(y) = \emptyset$, d'où $y \notin f(|s|)$. De plus l'ensemble $f(|s|)$ est fermé dans V. En effet $f_{||s|} : |s| \to V$ est une application propre entre sous-espaces localement compacts, donc l'image est fermée. Il en résulte que $s_y = 0$. Montrons que φ est surjectif. Comme Y est régulier on peut trouver une sous-famille $\mathcal{W}$ de la famille ordonnée $\mathcal{V}_y$ des voisinages ouverts de y, telle que $\bigcap_{V \in \mathcal{W}} \overline{V} = \{y\}$. On en déduit aisément que $\bigcap_{V \in \mathcal{W}} f^{-1}(V) = f^{-1}(y)$. Si $V \in \mathcal{W}$ considérons la famille de supports $\Omega_{f^{-1}(V)} = \{K \cap f^{-1}(V) \mid K \in c\}$. On a alors ([S],chap VII, exemple 2)

$$\Gamma_c\left(f^{-1}(y);A_{|f^{-1}(y)}\right) = \varinjlim_{V \in \mathcal{W}} \Gamma_{\Omega_{f^{-1}(V)}}(f^{-1}(V);A)$$

De plus on a $\Omega_{f^{-1}(V)} \subset \Phi_V$; en effet si $C \in \Omega_{f^{-1}(V)}$ alors $C = K \cap f^{-1}(V)$ où K est un compact de X, donc C est fermé dans $f^{-1}(V)$; et si K' est un compact de V alors $f^{-1}_{|C}(K') = K \cap f^{-1}(K')$ est compact; ainsi $C \in \Phi_V$. On a donc une inclusion $\Gamma_{\Omega_{f^{-1}(V)}}(f^{-1}(V);A) \to \Gamma_{\Phi_V}(f^{-1}(V);A)$.

Soit $t \in \Gamma_c(f^{-1}(y);A_{|f^{-1}(y)})$; soit $V \in \mathcal{W}$ et $s \in \Gamma_{\Omega_{f^{-1}(V)}}(f^{-1}(V);A)$ un représentant de t. Alors $s \in \Gamma_{\Phi_V}(f^{-1}(V);A)$.

Si s_y désigne l'image de s dans $\varinjlim_{V \in \mathcal{V}_y} \Gamma_{\Phi_V}(f^{-1}(V);A) = (f_!A)_y$, il est clair que $\varphi(s_y) = t$.

2.7 COROLLAIRE : $R^i(f_!A)_y = H^i_c(f^{-1}(y);A)$

2.8 On va établir maintenant quelques résultats qui seront utiles au paragraphe suivant. Rappelons pour commencer les deux définitions suivantes :

Un faisceau $\mathcal{F}$ sur X est $f_!$-acyclique si $R^i(f_!\mathcal{F}) = 0$ pour tout entier $i \geq 1$.

Le foncteur $f_!$ est de dimension cohomologique finie s'il existe un entier $n \geq 0$ tel que pour tout faisceau $\mathcal{F}$ sur X on a $R^i(f_!\mathcal{F}) = 0$ pour $i > n$.

Enfin il sera commode pour la suite de l'exposé de poser encore la définition suivante.

Un faisceau F sur X est $f_!$-mou si pour tout $y \in Y$ le faisceau $F_{|f^{-1}(y)}$ est c-mou.

2.9 *Exemples*

2.9.1 Dans le cas où Y est un point les définitions précédentes redonnent les définitions classiques de faisceau c-acyclique, d'espaces de c-dimension finie et de faisceau c-mou.

2.9.2 Si $j : X \to Y$ est l'inclusion d'un ouvert ou d'un fermé X dans Y alors $j_!$ est exact, donc tout faisceau est $j_!$-acyclique et $j_!$ est de dimension cohomologique nulle.

2.10 LEMME : *Soit F un faisceau sur X.*

1) F c-mou $\Rightarrow$ F $f_!$-mou $\Rightarrow$ F $f_!$-acyclique.

2) F est $f_!$-mou si et seulement si F_U est $f_!$-mou pour tout $U \in Ouv(X)$. (on rappelle que $F_U = (F_{|U})^X$).

3) Si F est $f_!$-mou alors F_U est $f_!$-acyclique pour tout $U \in Ouv(X)$.

Démonstration :

1) La première implication résulte de ([B],chap.II, prop.9.2). De plus le faisceau $F_{|f^{-1}(y)}$ est c-acyclique (idem, th.9.8); on a donc d'après le corollaire 2.7. $R^i(f_!F)_y = H^i_c(f^{-1}(y);F) = 0$ pour tout entier $i>0$ et tout $y \in Y$.

2) Si F_U est $f_!$-mou pour tout ouvert U de X alors il est évident que F est $f_!$-mou. Inversément si F est $f_!$-mou et si U est un ouvert de X on a $(F_U)_{|f^{-1}(y)} = \left(F_{|f^{-1}(y)}\right)_{U \cap f^{-1}(y)}$ et ce dernier faisceau est c-mou ([B], chap.II, coroll. 9.10).

3) C'est une conséquence de 1) et 2).

2.11 COROLLAIRE. *Si $\mathcal{I}$ est un faisceau injectif sur X alors $\mathcal{I}$ est $f_!$-mou et $\mathcal{I}_U$ est $f_!$-acyclique pour tout $U \in Ouv(X)$.*

Démonstration : En effet si $\mathcal{I}$ est injectif alors $\mathcal{I}$ est c-mou ([B],chap.II, prop.5.2 et corollaire 9.5).

2.12 THEOREME : *Si le foncteur $f_!$ est de dimension cohomologique finie alors tout faisceau $\mathcal{F}$ sur X admet une résolution bornée $0 \to \mathcal{F} \to \mathcal{L}^{\bullet}$ par des faisceaux qui sont $f_!$-mous.*

Démonstration : Supposons que la dimension cohomologique de $f_!$ est inférieure à n et soit $\mathcal{F}$ un faisceau sur X; on a donc $R^i(f_!\mathcal{F}) = 0$ pour tout $i>n$.
Considérons une résolution injective $0 \to \mathcal{F} \to \mathcal{I}^{\bullet}$ de $\mathcal{F}$ et posons $\mathcal{B} = \mathrm{Coker}(d_n : \mathcal{I}^n \to \mathcal{I}^{n+1})$. On obtient ainsi une résolution bornée de $\mathcal{F}$

$$(2.12.1) \qquad 0 \to \mathcal{F} \to \mathcal{I}^0 \to \mathcal{I}^1 \to \dots \to \mathcal{I}^n \to \mathcal{B} \to 0$$

dans laquelle les faisceaux $\mathcal{I}^p$ pour $p = 0,1,\dots,n$ sont $f_!$-mous en vertu de 2.11.

Il reste à montrer que $\mathcal{B}$ est aussi $f_!$-mou; pour cela il suffit de montrer que $H^1_c(f^{-1}(y);(\mathcal{B}|_{f^{-1}(y)})_V) = 0$ pour tout ouvert V de $f^{-1}(y)$ ([B],chap.II, prop.15.1). Mais V est un ouvert de $f^{-1}(y)$ si et seulement si il existe un ouvert U de X tel que $V = U\cap f^{-1}(y)$; de plus $\left(\mathcal{B}|_{f^{-1}(y)}\right)_{U\cap f^{-1}(y)} = (\mathcal{B}_U)|_{f^{-1}(y)}$. Il suffit donc de montrer que $H^1_c(f^{-1}(y);(\mathcal{B}_U)|_{f^{-1}(y)}) = 0$ pour tout ouvert U de X. Or si U est un ouvert de X, la suite

$$0 \to \mathcal{F}_U \to \mathcal{I}^0_U \to \mathcal{I}^1_U \to \dots \to \mathcal{I}^n_U \to \mathcal{B}_U \to 0$$

est exacte et les faisceaux $\mathcal{I}^p_U$, pour $0\leq p\leq n$, sont $f_!$-acycliques en vertu de 2.11. Il en résulte que $R^i(f_!\mathcal{B}_U) = R^{i+n}(f_!\mathcal{F}_U)$ pour $i\geq 1$ ([S],chap.III, lemme 6); ainsi d'après l'hypothèse sur la dimension de $f_!$ on en déduit que $R^i(f_!\mathcal{B}_U) = 0$ pour $i\geq 1$. Donc pour tout ouvert U de

X le faisceau $\mathcal{B}_U$ est $f_!$-acyclique. Mais on a $H^i_c(f^{-1}(y);(\mathcal{B}_U)_{|f^{-1}(y)}) = R^i(f_!\mathcal{B}_U)_y$ en vertu de 2.7; d'où la conclusion.

2.13 Pour faciliter la suite de l'exposé on va encore introduire les définitions suivantes qui sont suggérées par le théorème précédent.

Soit $\mathcal{F}$ un faisceau sur X.

On appelle $f_!$-résolution de $\mathcal{F}$ une résolution bornée de $\mathcal{F}$ par des faisceaux qui sont $f_!$-mous.

On appelle résolution de type standard de $\mathcal{F}$ une suite exacte

$$0 \to \mathcal{F} \to \mathcal{L}^0 \to \mathcal{L}^1 \to \dots \to \mathcal{L}^{p-1} \to \mathcal{L}^p \to 0$$

où les faisceaux $\mathcal{L}^j$ pour $j = 0,1,\dots,p-1$ sont injectifs et le faisceau $\mathcal{L}^p$ est $f_!$-mou.

En vertu du corollaire 2.11. une résolution de type standard de $\mathcal{F}$ est une $f_!$-résolution de $\mathcal{F}$.

De plus la démonstration du théorème 2.12 montre que si $f_!$ est de dimension cohomologique finie, tout faisceau admet une résolution de type standard.

2.14 PROPOSITION : *Soit $\mathcal{F}$ un faisceau sur X.*

1) Si $0 \to \mathcal{F} \to K^\cdot$ est une résolution bornée de $\mathcal{F}$ il existe une résolution de type standard $0 \to \mathcal{F} \to \mathcal{L}^\cdot$ et un morphisme de résolution $K^\cdot \to \mathcal{L}^\cdot$ au-dessus de $\mathcal{F}$.

2) Si $0 \to \mathcal{F} \to \mathcal{L}^\cdot$ et $0 \to \mathcal{F} \to M^\cdot$ sont deux résolutions de type standard il existe toujours un morphisme de résolution au-dessus de $\mathcal{F}$ de $\mathcal{L}^\cdot$ dans $M^\cdot$ ou de $M^\cdot$ dans $\mathcal{L}^\cdot$.

Démonstration :

1) Comme la catégorie Sh(X) admet suffisamment d'objets injectifs il existe une résolution injective $I^\cdot$ de $\mathcal{F}$ (en général non bornée) et un morphisme injectif $i : K^\cdot \to I^\cdot$ au-dessus de $\mathcal{F}$ ([H],chap.I,

lemme 4.6).

Supposons que la dimension cohomologique de $f_!$ est inférieure à n et soit q la longueur de la résolution $\mathcal{K}^\bullet$.

Si q≤n la solution est donnée par le diagramme commutatif et exact suivant

$$\begin{array}{ccccccccccccccccccc}
0 & \to & \mathcal{F} & \to & \mathcal{K}^0 & \to & \cdots & \to & \mathcal{K}^q & \to & 0 & \to & \cdots & \to & 0 & \to & 0 & \longrightarrow & 0 \\
 & & \| & & \downarrow{\scriptstyle i^0} & & & & \downarrow{\scriptstyle i^q} & & \downarrow & & & & \vdots & & \downarrow & & \\
0 & \to & \mathcal{F} & \to & \mathcal{I}^0 & \to & \cdots & \to & \mathcal{I}^q & \to & \mathcal{I}^{q+1} & \to & \cdots & \to & \mathcal{I}^n & \to & \operatorname{Coker} d_n & \to & 0
\end{array}$$

Si q≥n+1 la solution est donnée par le diagramme commutatif et exact suivant

$$\begin{array}{ccccccccccccccc}
0 & \to & \mathcal{F} & \to & \mathcal{K}^0 & \to & \cdots & \to & \mathcal{K}^{q-2} & \to & \mathcal{K}^{q-1} & \to & \mathcal{K}^q & \longrightarrow & 0 \\
 & & \| & & \downarrow{\scriptstyle i^0} & & & & \downarrow{\scriptstyle i^{q-2}} & & \downarrow{\scriptstyle i^{q-1}} & & \vdots{\scriptstyle j} & & \\
0 & \to & \mathcal{F} & \to & \mathcal{I}^0 & \to & \cdots & \to & \mathcal{I}^{q-2} & \to & \mathcal{I}^{q-1} & \to & \operatorname{Coker} d_{q-1} & \to & 0
\end{array}$$

où on a construit le morphisme j grâce à l'exactitude des lignes de ce diagramme.

2) Soit p et q la longueur des résolutions $\mathcal{L}^\bullet$ et $\mathcal{M}^\bullet$. Si p≤q alors il existe un morphisme $\varphi : \mathcal{L}^\bullet \to \mathcal{M}^\bullet$ au-dessus de $\mathcal{F}$. La construction de φ est adaptée de la construction classique ([T],chap. 5, prop.2.3).

<u>2.15</u> On va étudier maintenant le comportement du foncteur $f_!$ vis-à-vis des sommes directes infinies de faisceaux.

THEOREME : *Soit* $\{A_\alpha\}_{\alpha\in\Lambda}$ *une famille de faisceaux sur* X; *on a un isomorphisme canonique*

$$\bigoplus_{\alpha\in\Lambda} (f_! A_\alpha) = f_! (\bigoplus_{\alpha\in\Lambda} A_\alpha)$$

Démonstration : Il convient de rappeler que $\bigoplus_{\alpha\in\Lambda} A_\alpha$ désigne le faisceau engendré par le préfaisceau $U \to \bigoplus_{\alpha\in\Lambda} (\mathcal{A}_\alpha(U))$; c'est donc le faisceau des sections de l'espace étalé $L(\bigoplus_{\alpha\in\Lambda} \mathcal{A}_\alpha) = \coprod_{x\in X}(\bigoplus_{\alpha\in\Lambda} (\mathcal{A}_\alpha)_x)$.

Les morphismes de faisceaux évidents $\mathcal{A}_\alpha \to \bigoplus_{\alpha\in\Lambda} \mathcal{A}_\alpha$ induisent le morphisme canonique

$$\psi : \bigoplus_{\alpha\in\Lambda} (f_! \mathcal{A}_\alpha) \to f_!(\bigoplus_{\alpha\in\Lambda} \mathcal{A}_\alpha)$$

Pour montrer que ψ est un isomorphisme il suffit de montrer que, pour tout $y \in Y$, le morphisme induit sur les fibres

$$\psi_y : \bigoplus_{\alpha\in\Lambda} (f_! \mathcal{A}_\alpha)_y \to (f_!(\bigoplus_{\alpha\in\Lambda} \mathcal{A}_\alpha))_y$$

est un isomorphisme.

Compte tenu de l'isomorphisme de la proposition 2.6. et du fait que $(\bigoplus_{\alpha\in\Lambda} \mathcal{A}_\alpha)_{|f^{-1}(y)} = \bigoplus_{\alpha\in\Lambda} (\mathcal{A}_\alpha)_{|f^{-1}(y)}$, il suffit donc de montrer qu'on a un isomorphisme

$$\overline{\psi} : \bigoplus_{\alpha\in\Lambda} \Gamma_c(f^{-1}(y);(\mathcal{A}_\alpha)_{|f^{-1}(y)}) \to \Gamma_c(f^{-1}(y); \bigoplus_{\alpha\in\Lambda} ((\mathcal{A}_\alpha)_{|f^{-1}(y)})).$$

On est donc ramené à démontrer l'affirmation suivante :
soit $\{\mathcal{A}_\alpha\}_{\alpha\in\Lambda}$ une famille de faisceaux sur un espace compact X; alors le morphisme canonique

$$\psi : \bigoplus_{\alpha\in\Lambda} \Gamma(X;\mathcal{A}_\alpha) \to \Gamma(X; \bigoplus_{\alpha\in\Lambda} \mathcal{A}_\alpha)$$

est un isomorphisme.

Il est évident que ψ est injectif. Montrons donc que ψ est surjectif.

Une section continue $s : X \to L(\bigoplus_{\alpha\in\Lambda} \mathcal{A}_\alpha)$ donne naissance à des sections continues $s_\alpha : X \to L(\mathcal{A}_\alpha)$ définies en posant $s_\alpha(x) = p_\alpha(s(x))$ où p_α est l'application continue induite sur les espaces étalés par la projection canonique $\bigoplus_{\alpha\in\Lambda} \mathcal{A}_\alpha \to \mathcal{A}_\alpha$ ([T],chap.II,3.8).

Si $x \in X$ on a $s(x) = (s_\alpha(x))_{\alpha\in\Lambda}$ et il n'y a qu'un nombre fini d'indices α pour lesquels $s_\alpha(x) \neq 0$; de plus si $s_\alpha(x) = 0$ il y a un

petit voisinage ouvert de x sur lequel s_α reste nulle.
Ainsi pour chaque point $x \in X$ il existe un voisinage ouvert V_x de x sur lequel le nombre de composantes non nulles de s ne peut pas augmenter. La famille $\{V_x\}_{x \in X}$ recouvre X (il suffit d'ailleurs de recouvrir le support de s); on peut donc en extraire un recouvrement fini; par suite il n'y a qu'un nombre fini de sections s_α qui ne sont pas identiquement nulles. Donc $(s_\alpha)_{\alpha\in\Lambda} \in \bigoplus_{\alpha\in\Lambda}\Gamma(X;A_\alpha)$ et $\psi((s_\alpha)_{\alpha\in\Lambda}) = s$.

2.16 Le théorème 2.15 est faux pour le foncteur f_*. Prenons $X = \bigcup_{n\in\mathbb{N}} U_n$, où $U_n =]2n\ 2n+1[$;muni de la topologie induite par celle de $\mathbb{R}$ l'espace X est localement compact. Prenons $A_n = \mathbb{Z}_{U_n}$ le faisceau constant sur U_n, étendu par zéro en dehors de U_n. Comme la famille $\{A_n\}_{n\in\mathbb{N}}$ est localement finie, le préfaisceau $U \to \bigoplus_{n\in\mathbb{N}} A_n(U)$ est un faisceau.

Considérons alors la section $s \in \Gamma(X; \bigoplus_{n\in\mathbb{N}} A_n)$ qui est définie par la propriété suivante : pour tout $x \in X$ et tout $n \in \mathbb{N}$,

$$p_n(s(x)) = \begin{cases} 1 \text{ si } x \in U_n \\ 0 \text{ si } x \notin U_n \end{cases} .$$

Il est évident que cette section ne peut pas s'écrire comme une somme directe $\bigoplus_{n\in\mathbb{N}} s_n$ avec seulement un nombre fini de s_n non identiquement nulle. Donc $\Gamma(X; \bigoplus_{n\in\mathbb{N}} A_n) \neq \bigoplus_{n\in\mathbb{N}} \Gamma(X;A_n)$.

2.17 PROPOSITION : *Le foncteur $f_!$ est exact sur la catégorie Inj(X).*

Démonstration : Si $0 \to A \to B \to C \to 0$ est une suite exacte de faisceaux injectifs sur X on a $B = A \oplus C$; il en résulte que $f_! B = f_! A \oplus f_! C$ et par conséquent la suite $0 \to f_! A \to f_! B \to f_! C \to 0$ est exacte.

3. LE FONCTEUR $f^{!}$

3.1 Si X est un espace topologique nous désignerons par $C^{+}(X)$ la catégorie dont les objets sont les complexes bornés inférieurement de faisceaux de A-modules sur X et dont les flèches sont les morphismes de complexes. Nous désignerons encore par $K^{+}(X)$ la catégorie dont les objets sont ceux de $C^{+}(X)$ et dont les flèches sont les classes d'homotopie de morphismes de complexes. Nous désignerons enfin par $I^{+}(X)$ la sous-catégorie pleine de $K^{+}(X)$ dont les objets sont les faisceaux de A-modules injectifs sur X.
La catégorie dérivée associée sera notée $D^{+}(X)$ (voir [H], chap.I).

3.2 Soit X et Y deux espaces localement compacts. Considérons une application continue $f : X \to Y$ telle que le foncteur $f_!$ soit de dimension cohomologique finie. Notons $\underline{\underline{A}}$ le faisceau constant sur X qui admet l'anneau commutatif unitaire A comme fibre. D'après le théorème 2.12, on peut trouver une $f_!$-résolution $0 \to \underline{\underline{A}} \to K^{\cdot}$ du faisceau $\underline{\underline{A}}$. On va tout d'abord définir un foncteur

$$f^{!}_{K^{\cdot}} : K^{+}(Y) \to K^{+}(X)$$

On verra ensuite que ce foncteur induit sur les catégories dérivées un foncteur qui est indépendant du choix de $K^{\cdot}$.

3.3 Soit $G^{\cdot}$ un objet de $K^{+}(Y)$; sans restreindre la généralité on peut supposer que $G^{i} = 0$ si $i<0$. Pour tout $U \in \mathrm{Ouv}(X)$ posons

$$f^{!}_{K^{\cdot}}(G^{\cdot})(U) = \mathrm{Hom}^{\cdot}(f_!(K^{\cdot}_U);G^{\cdot})$$

Maintenant soit $U,V \in \mathrm{Ouv}(X)$ tels que $U \subset V$. On définit tout d'abord un morphisme injectif $r_{U,V} : K^{\cdot}_U \to K^{\cdot}_V$ de complexes de faisceaux sur X de la façon suivante : soit $W \in \mathrm{Ouv}(X)$; si $s \in \Gamma(W;K^{\cdot}_U)$ alors par définition $s \in \Gamma(W \cap U;K^{\cdot})$ et le support de s est fermé dans W; on peut donc étendre s par zéro sur $W \cap V$ de façon à obtenir une section $\tilde{s} \in \Gamma(W;K^{\cdot}_V)$; on pose alors $r_{U,V}(W)(s) = \tilde{s}$.

On peut maintenant définir des morphismes

$$\rho_{U,V} : f^{!}_{K^\cdot}(G^\cdot)(V) \to f^{!}_{K^\cdot}(G^\cdot)(U)$$

en posant

$$\rho_{U,V}(\varphi) = \varphi\circ f_{!}(r_{U,V}) \text{ si } \varphi\in \mathrm{Hom}^\cdot(f_{!}(K^\cdot_V);G^\cdot).$$

Enfin on définit une différentielle de degré +1

$$\partial_U : f^{!}_{K^\cdot}(G^\cdot)(U) \to f^{!}_{K^\cdot}(G^\cdot)(U)$$

en posant

$$\partial_U(\varphi) = d'\circ\varphi-(-1)^{\deg(\varphi)}\varphi\circ d'' \text{ si } \varphi\in \mathrm{Hom}^\cdot(f_{!}(K^\cdot_U);G^\cdot),$$

où d' est la différentielle du complexe $G^\cdot$ et d" est la différentielle induite sur $f_{!}(K^\cdot_U)$ par la différentielle du complexe $K^\cdot$.

3.4 LEMME : *La correspondance* $U \to f^{!}_{K^\cdot}(G^\cdot)(U)$, *avec les morphismes* $\rho_{U,V}$ *et les différentielles* ∂_U, *définit un complexe borné inférieurement de préfaisceaux sur X. Ce complexe sera noté* $f^{!}_{K^\cdot}(G^\cdot)$.

Démonstration : Il est immédiat que les morphismes $\rho_{U,V}$ satisfont les conditions de compatibilité des morphismes de restriction. De plus les ∂_U sont compatibles avec ces restrictions donc définissent une différentielle ∂ sur $f^{!}_{K^\cdot}(G^\cdot)$ de degré +1. Enfin la composante de degré $k\in\mathbb{Z}$ du complexe de A-modules $f^{!}_{K^\cdot}(G^\cdot)(U)$ est donnée par $\prod_{i\in\mathbb{Z}} \mathrm{Hom}(f_{!}(K^i_U);G^{i+k})$; comme le complexe $K^\cdot_U$ est doublement borné et le complexe $G^\cdot$ est borné inférieurement, ceci montre que le complexe $f^{!}_{K^\cdot}(G^\cdot)$ est borné inférieurement.

3.5 THEOREME : *Si* $G^\cdot$ *est un objet de* $K^+(Y)$ *alors* $f^{!}_{K^\cdot}(G^\cdot)$ *est un objet de* $K^+(X)$.

Démonstration : Soit $\{U_\alpha\}_{\alpha\in\Lambda}$ une famille d'ouverts de X ; posons $U = \bigcup_{\alpha\in\Lambda} U_\alpha$ et $U_{\alpha\beta} = U_\alpha \cap U_\beta$ et considérons la suite

(3.5.1)
$$0 \to f^!_{K^\cdot}(G^\cdot)(U) \xrightarrow{\varphi} \prod_{\alpha\in\Lambda} f^!_{K^\cdot}(G^\cdot)(U_\alpha) \xrightarrow{\psi} \prod_{(\alpha,\beta)\in\Lambda\times\Lambda} f^!_{K^\cdot}(G^\cdot)(U_{\alpha\beta})$$

où
$$\varphi(s) = (\rho_{U_\alpha,U}(s))_{\alpha\in\Lambda}$$

et
$$\psi((s_\alpha)_{\alpha\in\Lambda}) = (\rho_{U_{\alpha\beta},U_\alpha}(s_\alpha) - \rho_{U_{\alpha\beta},U_\beta}(s_\beta))_{(\alpha,\beta)\in\Lambda\times\Lambda}$$

Il faut démontrer que cette suite est exacte.
En tenant compte des propriétés du foncteur $\mathrm{Hom}^\cdot$ vis-à-vis du produit direct et de la somme directe et en utilisant le fait que le foncteur contravariant $\mathrm{Hom}^\cdot(-;G^\cdot)$ est exact gauche, on voit que l'exactitude de la suite (3.5.1) est impliquée par l'exactitude de la suite

(3.5.2)
$$\bigoplus_{(\alpha,\beta)\in\Lambda\times\Lambda} f_!(K^\cdot_{U_{\alpha\beta}}) \to \bigoplus_{\alpha\in\Lambda} f_!(K^\cdot_{U_\alpha}) \to f_!(K^\cdot_U) \to 0$$

Mais d'après le théorème 2.15 l'exactitude de cette suite est équivalente à l'exactitude de la suite

(3.5.3)
$$f_!\Big(\bigoplus_{(\alpha,\beta)\in\Lambda\times\Lambda} K^\cdot_{U_{\alpha\beta}}\Big) \to f_!\Big(\bigoplus_{\alpha\in\Lambda} K^\cdot_{U_\alpha}\Big) \to f_!(K^\cdot_U) \to 0.$$

Considérons donc la suite de faisceaux sur X

(3.5.4)
$$\bigoplus_{(\alpha,\beta)\in\Lambda\times\Lambda} K^\cdot_{U_{\alpha\beta}} \xrightarrow{\bar\varphi} \bigoplus_{\alpha\in\Lambda} K^\cdot_{U_\alpha} \xrightarrow{\bar\psi} K^\cdot_U \to 0$$

où $\bar\psi$ est le morphisme canonique induit par les morphismes $r_{U_\alpha,U} : K^\cdot_{U_\alpha} \to K^\cdot_U$ et $\bar\varphi$ est le morphisme canonique induit par les morphismes $\ell_\alpha \circ r_{U_{\alpha\beta},U_\alpha} - \ell_\beta \circ r_{U_{\alpha\beta},U_\beta} : K^\cdot_{U_{\alpha\beta}} \to \bigoplus_{\alpha\in\Lambda} K^\cdot_{U_\alpha}$, ℓ_γ désignant l'inclusion canonique de $K^\cdot_{U_\gamma}$ dans $\bigoplus_{\alpha\in\Lambda} K^\cdot_{U_\alpha}$. La suite (3.5.4) est exacte. En effet soit $x \in X$ et considérons la suite des fibres

(3.5.5)
$$\bigoplus_{(\alpha,\beta)\in\Lambda\times\Lambda} (K^\cdot_{U_{\alpha\beta}})_x \xrightarrow{\bar\varphi_x} \bigoplus_{\alpha\in\Lambda} (K^\cdot_{U_\alpha})_x \xrightarrow{\bar\psi_x} (K^\cdot_U)_x \to 0$$

Si $x \notin U$ tous les termes de cette suite sont nuls. Sinon posons $\Lambda_x = \{\alpha \in \Lambda | U_\alpha \ni x\}$; on a alors la suite

$$\bigoplus_{(\alpha,\beta)\in\Lambda_x\times\Lambda_x} K^\bullet_x \xrightarrow{\overline{\varphi}_x} \bigoplus_{\alpha\in\Lambda_x} K^\bullet_x \xrightarrow{\overline{\psi}_x} K^\bullet_x \longrightarrow 0$$

où $\overline{\psi}_x$ est le morphisme canonique induit par le morphisme identité de $K^\bullet_x$ et $\overline{\varphi}_x$ est le morphisme canonique induit par les morphismes $\ell_\alpha - \ell_\beta : K^\bullet_x \to \bigoplus_{\alpha\in\Lambda_x} K^\bullet_x$; cette suite est clairement exacte.

Pour terminer la démonstration il reste à prouver que le foncteur $f_!$ préserve l'exactitude de la suite (3.5.4).

Pour chaque entier $n \geq 1$ posons $U_{(\alpha_1;\ldots;\alpha_n)} = U_{\alpha_1} \cap \ldots \cap U_{\alpha_n}$ et considérons la suite de faisceaux sur X

(3.5.6)

$$\ldots \to \bigoplus_{(\alpha_1;\ldots;\alpha_n)\in\Lambda^n} K^\bullet_{U_{(\alpha_1;\ldots;\alpha_n)}} \to \ldots \to \bigoplus_{(\alpha_1;\alpha_2)\in\Lambda^2} K^\bullet_{U_{(\alpha_1;\alpha_2)}} \to \bigoplus_{\alpha\in\Lambda} K^\bullet_{U_\alpha} \to K^\bullet_U \to 0$$

qui prolonge la suite (3.5.4). Les flèches de cette suite sont les morphismes canoniques induits comme précédemment par les morphismes

$$\sum_{i=1}^{n} (-1)^i \ell_{(\alpha_1;\ldots;\hat{\alpha}_i;\ldots;\alpha_n)} \circ r_{U_{(\alpha_1;\ldots;\alpha_n)}, U_{(\alpha_1;\ldots;\hat{\alpha}_i;\ldots;\alpha_n)}}$$

$$: K^\bullet_{U_{(\alpha_1;\ldots;\alpha_n)}} \to \bigoplus_{(\alpha_1;\ldots;\alpha_{n-1})\in\Lambda^{n-1}} K^\bullet_{U_{(\alpha_1;\ldots;\alpha_{n-1})}}$$

Une démonstration analogue à la précédente montre que (3.5.6) est une suite exacte, et les faisceaux des complexes de cette suite sont $f_!$-acycliques d'après 2.10.3.

Comme le foncteur $f_!$ est exact gauche (2.3), exact sur les faisceaux injectifs (2.17) et de dimension cohomologique finie par hypothèse, il résulte du corollaire A.3 de l'appendice que $f_!$ préserve l'exactitude de la suite (3.5.6), donc de la suite (3.5.4).

3.6 COROLLAIRE : *La correspondance* $G^\cdot \mapsto f^!_{K^\cdot}(G^\cdot)$ *définit un foncteur* $f^!_{K^\cdot} : K^+(Y) \to K^+(X)$.
De plus la correspondance $K^\cdot \mapsto f^!_{K^\cdot}$ *définit un foncteur contravariant de la catégorie des* $f_!$*-résolutions de* $\underline{\underline{A}}$ *et des morphismes de résolution au-dessus de* $\underline{\underline{A}}$ *dans la catégorie des foncteurs de* $K^+(Y)$ *dans* $K^+(X)$ *et des morphismes de foncteurs.*

3.7 PROPOSITION : *Si* $G^\cdot \in ObI^+(Y)$ *alors le faisceau* $f^!_{K^\cdot}(G^\cdot)$ *est flasque.*

Démonstration : Si U est un ouvert de X on a une injection canonique

$$0 \to K^\cdot_U \to K^\cdot$$

d'où une injection

$$0 \to f_! K^\cdot_U \to f_! K^\cdot$$

puisque $f_!$ est exact gauche. Mais le foncteur $Hom^\cdot(-;G^\cdot)$ est exact puisque $G^\cdot$ est injectif. On en déduit que le morphisme canonique

$$f^!_{K^\cdot}(G^\cdot)(X) \to f^!_{K^\cdot}(G^\cdot)(U) \to 0$$

est surjectif.

3.8 THEOREME : *Soit* $G^\cdot \in ObI^+(Y)$. *Si* $K^\cdot$ *et* $L^\cdot$ *sont deux* $f_!$*-résolutions du faisceau* $\underline{\underline{A}}$, *alors les complexes de faisceaux* $f^!_{K^\cdot}(G^\cdot)$ *et* $f^!_{L^\cdot}(G^\cdot)$ *sont quasi-isomorphes.*

Démonstration : D'après la proposition 2.14. 1) il existe des résolutions de type standard $M^\cdot$ et $N^\cdot$ et des morphismes de résolutions $K^\cdot \to M^\cdot$ et $L^\cdot \to N^\cdot$ au-dessus de $\underline{\underline{A}}$. De plus par 2.14 2) on peut trouver un morphisme de résolution $M^\cdot \to N^\cdot$ ou $N^\cdot \to M^\cdot$ au-dessus de $\underline{\underline{A}}$.
Le théorème est alors une conséquence immédiate du résultat suivant.

3.9 LEMME : *Soit* $K^\cdot$ *et* $L^\cdot$ *deux* $f_!$*-résolutions de* $\underline{\underline{A}}$. *Si* $\varphi : K^\cdot \to L^\cdot$ *est un morphisme de résolutions au-dessus de* $\underline{\underline{A}}$ *alors le* *morphisme* $\overline{\psi} : f^{!}_{L^\cdot} \to f^{!}_{K^\cdot}$ *induit un quasi-isomorphisme* $\overline{\psi}(G^\cdot) : f^{!}_{L^\cdot}(G^\cdot) \to f^{!}_{K^\cdot}(G^\cdot)$ *pour tout* $G^\cdot \in ObI^+(Y)$.

Démonstration : Pour tout ouvert U de X le morphisme induit $\psi_U : K^\cdot_U \to L^\cdot_U$ est un quasi-isomorphisme car le foncteur $U \to F_U$ étant exact, il transforme résolutions en résolutions. Maintenant on a $H^q R^p(f_! K^\cdot_U) = H^q R^p(f_! L^\cdot_U) = 0$ pour tout q et tout $p \neq 0$ car les faisceaux K^q_U et L^q_U sont $f_!$-acycliques pour tout q (lemme 2.10.3)). Il en résulte que $f_!\psi_U : f_! K^\cdot_U \to f_! L^\cdot_U$ est un quasi-isomorphisme ([S],chap.IX, corollaire Prop.4).

Maintenant soit $G^\cdot \in ObI^+(Y)$ et $U \in Ouv(X)$; considérons le diagramme commutatif suivant

$$\begin{array}{ccccccccc}
0 \to & \mathrm{Ext}^{n-1}(H(f_!(L^\cdot_U));H(G^\cdot)) & \to & H^n(\mathrm{Hom}^\cdot(f_!(L^\cdot_U);G^\cdot)) & \to & \mathrm{Hom}^n(H(f_!(L^\cdot_U));H(G^\cdot)) & \to 0 \\
& \downarrow & & \downarrow {\scriptstyle \overline{\psi}(G^\cdot)(U)*} & & \downarrow & \\
0 \to & \mathrm{Ext}^{n-1}(H(f_!(K^\cdot_U));H(G^\cdot)) & \to & H^n(\mathrm{Hom}^\cdot(f_!(K^\cdot_U);G^\cdot)) & \to & \mathrm{Hom}^n(H(f_!(K^\cdot_U));H(G^\cdot)) & \to 0
\end{array}$$

Les lignes sont exactes car les faisceaux du complexe $G^\cdot$ sont injectifs ([G],chap.I, th.5.4.2); les deux flèches verticales extrêmes sont des isomorphismes car elles sont induites par l'isomorphisme $(f_!\psi_U)^* : H(f_!(K^\cdot_U)) \to H(f_!(L^\cdot_U))$.

Du lemme des cinq il résulte donc que la flèche $\overline{\psi}(G^\cdot)(U)*$ est un isomorphisme. Par suite $\overline{\psi}(G^\cdot)$ est un quasi-isomorphisme.

3.10 COROLLAIRE : *Le foncteur*

$$f^{!} : D^+(Y) \to D^+(X)$$

induit sur les catégories dérivées par le foncteur $f^{!}_{K^\cdot}$, *est indépendant, à isomorphisme près, du choix de la résolution* $K^\cdot$.

Démonstration : En effet si $K^\cdot$ et $L^\cdot$ sont deux $f_!$-résolutions de $\underline{\underline{A}}$ et si $P_X : K^+(X) \to D^+(X)$ est le foncteur canonique, il résulte du théorème 3.8 que les foncteurs $P_X \circ f^{!}_{K^\cdot}$ et $P_X \circ f^{!}_{L^\cdot}$ sont canoniquement isomorphes. De plus tout foncteur de $I^+(Y)$ dans $D^+(X)$ se factorise à travers $D^+(Y)$ ([H], chap.I). On obtient ainsi le foncteur $f^! : D^+(Y) \to D^+(X)$.

3.11 *Exemples*

3.11.1 Supposons que Y est un point et que X est un espace localement compact de c-dimension finie (2.9.1). Si $I^\cdot$ et $J^\cdot$ sont deux résolutions injectives de A il est évident que les faisceaux $f^!_{K^\cdot}(I^\cdot)$ et $f^!_{K^\cdot}(J^\cdot)$ sont quasi-isomorphes.
Pour tout ouvert U de X on pose

$$\mathbb{D}_X(U) = f^!_{K^\cdot}(I^\cdot)(U) = \mathrm{Hom}^\cdot(\Gamma_c(U;K^\cdot_U);I^\cdot).$$

D'après ce qui précède le faisceau $\mathbb{D}_X$ est bien défini sur la catégorie dérivée $D^+(X)$; on appelle $\mathbb{D}_X$ le faisceau dualisant sur X (pour plus de détails sur ce faisceau voir [GMP] et [Bo]).

3.11.2 Si $j : X \to Y$ est l'inclusion d'un sous-espace ouvert ou fermé X de Y, alors le foncteur $j^!$ est adjoint à droite au foncteur $j_!$.
On doit montrer que si $A^\cdot \in \mathrm{Ob}K^+(X)$ et $G^\cdot \in \mathrm{Ob}I^+(Y)$ on a un isomorphisme

$$\Phi : \mathrm{Hom}^\cdot_{K^+(Y)}(j_!(A^\cdot);G^\cdot) \to \mathrm{Hom}^\cdot_{K^+(X)}(A^\cdot;j^!(G^\cdot))$$

D'après 2.5.2 on a $j_!(A^\cdot) = A^{\cdot Y}$.
D'autre part, d'après 2.9.2, le foncteur $j_!$ est de dimension cohomologique nulle, donc $0 \to \underline{\underline{A}} \to \underline{\underline{A}} \to 0$ est une $j_!$-résolution du faisceau constant $\underline{\underline{A}}$ sur X. Supposons tout d'abord que X est un sous-espace ouvert de Y et soit $U \in \mathrm{Ouv}(X)$. On a alors $(\underline{\underline{A}}_U)^Y = \underline{\underline{A}}_U$ (en considérant dans le membre de droite $\underline{\underline{A}}$ comme le faisceau constant sur Y), d'où on en déduit, compte tenu de ([G],chap.II, Remarque 2.9.1) et de 1.6,

$$j^!(G^\cdot)(U) = \mathrm{Hom}^\cdot(\underline{\underline{A}}_U;G^\cdot) = \Gamma(U;G^\cdot) = \Gamma(U;j^*G^\cdot)$$

Donc si X est un ouvert de Y on a $j^! = j^*$.
L'isomorphisme Φ est alors une conséquence de ([T],chap.III, Remark 8.13).

Supposons maintenant que X est un sous-espace fermé de Y. Soit $U \in \mathrm{Ouv}(X)$ et choisissons $V \in \mathrm{Ouv}(Y)$ tel que $U = V \cap X$. On a alors $(\underline{\underline{A}}_U)^Y = (\underline{\underline{A}}_X)_V$, d'où on en déduit

$$j^{!}(G^{\cdot})(U) = \mathrm{Hom}^{\cdot}((\underline{\underline{A}}_X)_V;G^{\cdot}) = \Gamma_{(X)}(V;G^{\cdot})$$

où (X) désigne la famille des ensembles fermés de Y qui sont contenus dans X.

Rappelons que, puisque X est un fermé de Y, on a $j_! = j_*$ (2.5.2).

On peut maintenant construire le morphisme Φ.

Soit $\varphi \in \mathrm{Hom}^{\cdot}(j_*(A^{\cdot});G^{\cdot})$. Soit $U \in \mathrm{Ouv}(X)$ et soit $V \in \mathrm{Ouv}(Y)$ tel que $U = V \cap X$. Il faut définir un morphisme $\overline{\varphi}(U) : \Gamma(U;A^{\cdot}) \to \Gamma_{(X)}(V;G^{\cdot})$ compatible avec les restrictions. On a $\Gamma(U;A^{\cdot}) = \Gamma(V \cap X;A^{\cdot}) =$ $= \Gamma(V;j_*A)$. Si $s \in \Gamma(V;j_*A)$ on a $|\varphi(V)(s)| \subset X$; en effet si $y \in Y-X$ on a $(\varphi(V)(s))_y = \varphi_y(s_y) = 0$. On définit alors $\overline{\varphi}(U)$ en posant $\overline{\varphi}(U)(s) = \varphi(V)(s)$ et on définit Φ en posant $\Phi(\varphi) = \overline{\varphi}$.

Le morphisme réciproque de Φ est immédiat à construire.

On en déduit que Φ est un isomorphisme.

On a donc des flèches d'adjonction

$$\alpha : j_! j^! \to 1_{K^+(Y)}$$

$$\beta : 1_{K^+(X)} \to j^! j_!$$

Enfin un argument classique déjà utilisé montre que $j^!(G^{\cdot}) \in \mathrm{ObI}^+(X)$.

Appendice

A.1 Soit $\underline{C}$ et $\underline{D}$ deux catégories abéliennes; on suppose que $\underline{C}$ admet suffisamment d'objets injectifs. Soit $F : \underline{C} \to \underline{D}$ un foncteur exact gauche, exact sur les objets injectifs et de dimension cohomologique finie.

A.2 PROPOSITION : *Soit*

$$\dots \to A_3 \xrightarrow{d_3} A_2 \xrightarrow{d_2} A_1 \xrightarrow{d_1} A_0 \xrightarrow{d_0} 0$$

une suite exacte dans $\underline{C}$ telle que les objets A_p, pour $p \geq 0$, sont F-acycliques. Alors les noyaux $Z_p = \mathrm{Ker}\, d_p$ sont aussi F-acycliques.

Démonstration : Pour chaque entier $p \geq 0$ on a une suite exacte

$$0 \to Z_{p+1} \to A_{p+1} \to Z_p \to 0.$$

On peut trouver des résolutions injectives $0 \to Z_j \to Y_j^{\cdot}$ et $0 \to A_j \to X_j^{\cdot}$ telles que chacune des suites

$$0 \to Y_{p+1}^{\cdot} \to X_{p+1}^{\cdot} \to Y_p^{\cdot} \to 0$$

soit exacte. D'après l'hypothèse faite sur F chacune des suites

$$0 \to F(Y_{p+1}^{\cdot}) \to F(X_{p+1}^{\cdot}) \to F(Y_p^{\cdot}) \to 0$$

est encore exacte.

Comme les complexes $F(X_j^{\cdot})$ sont acycliques, il résulte de la suite exacte longue de cohomologie que l'on a pour tout entier $i \geq 1$ et $p \geq 1$ des isomorphismes $H^i(F(Y_p^{\cdot})) = H^{i+1}(F(Y_{p+1}^{\cdot}))$; donc $H^i(F(Y_p^{\cdot})) = H^{i+q}(F(Y_{p+q}^{\cdot}))$ pour tout entier $q \geq 1$. Comme F est de dimension cohomologique finie on a $H^{i+q}(F(Y_{p+q}^{\cdot})) = 0$ dès que q est assez grand. Il en résulte que $H^i(F(Y_p^{\cdot})) = 0$ pour tout $i \geq 1$; donc Z_p est F-acyclique.

A.3. COROLLAIRE : *Soit*

$$\dots \to A_3 \to A_2 \to A_1 \to A_0 \to 0$$

une suite exacte dont chacun des termes est F-acyclique. Alors la suite

$$\dots \to F(A_3) \to F(A_2) \to F(A_1) \to F(A_0) \to 0$$

est encore exacte.

Démonstration : Il suffit d'appliquer ([S],chap.III, prop.7) aux suites exactes

$$0 \to Z_p \to A_p \to A_{p-1} \to \dots \to A_1 \to A_0 \to 0.$$

BIBLIOGRAPHIE

[B] G.E. Bredon : "Sheaf theory" New York, McGraw-Hill, 1967,(McGraw Hill series in higher mathematics).

[Bo] A. Borel et al. "Seminar on intersection homology", preprint E.T.H. Zürich, 1983.

[G] R.Godement : "Topologie algébrique et théorie des faisceaux", 2ème tirage, Paris, Hermann, 1964.

[GMP] M.Goresky - R. MacPherson : "Intersection homology II", Invent. math., 72(1983), 77-129.

[H] R. Hartshorne : "Residues and duality : lecture notes of a Seminar on the work of A.Grothendieck given at Harvard, 1963/64, Berlin, Springer, 1966, (Lecture notes in mathematics, vol. 20).

[S] R.G. Swan : "The theory of sheaves", Chicago, The University of Chicago press, 1964, (Chicago lectures in mathematics).

[T] B.R. Tennison : "Sheaf theory", Cambridge, Cambridge university press, 1975, (London mathematical society lecture notes series, vol.20).

[V] J.L. Verdier : "Dualité dans la cohomologie des espaces localement compacts", exposés No 300, Séminaire Bourbaki, vol. 1965/1966, exposés No 295-312, New York, W.A. Benjamin, 1966.

Intersection Cohomology
Progress in Mathematics, Vol. 50
© 1984 Birkhäuser Boston, Inc.

VII. WITT SPACE COBORDISM THEORY (after P. Siegel)

by M. Goresky

This is a report on the paper [S] of Paul Siegel.

1. *Motivation.* It was shown in [GM1] that if X is a compact piecewise linear pseudomanifold with even codimension strata, and if $\bar{m}$ represents the "middle perversity" then $IH_*^{\bar{m}}(X)$ satisfies Poincaré duality over the rationals. This gives rise to a cobordism invariant signature for 4k-dimensional pseudomanifolds (where the cobordisms are also required to have even codimension singularities). It then seems an interesting question to compute the cobordism groups of spaces having even codimension singularities. This turns out to be a bad question because a stratified space X might not be cobordant to the space X' which is obtained from X by refining the stratification. Therefore we must allow the spaces and cobordisms to have some strata of odd codimension, provided that these strata do not destroy the Poincaré duality of the intersection homology groups. Siegel found a natural class of such spaces (which he called Witt spaces), and calculated their cobordism groups by inventing an important surgery technique for singular spaces. These cobordism groups Ω_i^W turn out to be 0 unless $i \equiv 0$ (mod 4), and to coincide with the Witt group, W(Q) for i = 4k .[MH]

2. *Witt spaces.* An oriented piecewise linear pseudomanifold X is called a Witt space if, for each stratum S with odd codimension,

$$IH_{\ell/2}^{m}(L;Q) = 0 \qquad (*)$$

where L is the link of the stratum, $\ell = \dim(L) = \operatorname{codim}(S) - 1$ and $\bar{m}$ is "lower middle perversity", $m(c) = [(c - 2)/2]$.

If X is a Witt space with respect to one stratification, then it is also a Witt space with respect to any other stratification (this follows from the stratification invariance and topological invariance of the sheaf of intersection homology chains [GM2]).

THEOREM . If X is a compact Witt space, then $IH_*^{\bar{m}}(X;Q)$ satisfies Poincaré duality over the rationals, i.e. the intersection pairing

$$IH_i^{\bar{m}}(X;Q) \times IH_{n-i}^{\bar{m}}(X;Q) \to Q$$

is nondegenerate. (Here n = dim(X)).

Proof. Let $\bar{n}$ denote the "upper middle perversity", n(c) = [(c-1)/2]. Then $IH_*^{\bar{m}}(X;Q)$ is dual to $IH_*^{\bar{n}}(X;Q)$. However for a Witt space X, the natural map $IH_*^{\bar{m}}(X;Q) \to IH_*^{\bar{n}}(X;Q)$ is an isomorphism. This is almost a tautology from Deligne's construction of the sheaf of intersection chains.

We define Ω_i^W to be the cobordism group of i-dimensional Witt spaces. Thus Ω_i^W is generated by compact oriented i-dimensional Witt spaces (with additive structure given by connected sum, and action of -1 given by reversal of orientation) and has the relations $X \sim 0$ if $X = \partial Y$ where Y is an $n+1$-dimensional Witt space with collared boundary. (This means that each stratum of odd codimension in $Y - \partial Y$ must satisfy the condition (*)).

3. *When is* X *a boundary* ? First we notice that if i is odd, then $\Omega_i^W = 0$, because any i-dimensional Witt space X is the boundary of Y = cone(X). (Y is also a Witt space because the cone vertex is a stratum with even codimension in Y) . Similarly, if X is a 2k-dimensional Witt space and if $IH_k(X;Q) = 0$ then X represents the 0 element in Ω_i^W because Y = cone(X) is again a Witt space.

4. *The Witt group* W(Q). (see [MH]). W(Q) is the abelian group which is generated by rational vector-spaces with a symmetric nondegenerate bilinear pairing, $\beta : V \times V \to Q$, and has the relation $(V,\beta) = 0$ if V contains a self annihilating subspace $W = W^{\perp}$ such that $\dim(W) = \dim(V)/2$. The additive structure on W(Q) is given by the perpendicular direct sum of vector-spaces. This group has been calculated :

$$W(Q) \cong W(Z) \oplus \bigoplus_{p \text{ prime}} W(Z/(p))$$

where $W(Z) \cong Z$ and is given by the signature

$$W(Z/(2)) \cong Z/(2)$$

$$W(Z/(p)) \cong Z/(4) \text{ or } Z/(2) \oplus Z/(2) \quad \text{if } p = 2$$

To any 4k-dimensional Witt space X we have an associated rational vector-space and symmetric bilinear form,

$$[X] = (IH^{\bar{m}}_{2k}(X;Q), \text{ intersection pairing})$$

5. *The Main Theorem* . THEOREM. [S] The association $X \to [X]$ induces an isomorphism

$$w : \Omega^{W}_{4k} \to W(Q)$$

Furthermore, $\Omega^{W}_{i} = 0$ if $i \not\equiv 0 \pmod 4$.

Proof. First we check that w is well defined. If $X = \partial Y$ then the image of the boundary homomorphism

$$IH^{\bar{m}}_{2k+1}(Y,\partial Y) \to IH^{\bar{m}}_{2k}(X)$$

is a self annihilating subspace of half the dimension. Thus, $w(X) = 0$. Next, Siegel shows w is surjective by constructing for each element $\alpha \in W(Q)$ a specific 4-dimensional space which represents α in the Witt group. The interesting argument concerns the injectivity of w . Suppose that X is a connected 4k-dimensional Witt space such that $w(X) = 0$, i.e. $IH^{\bar{m}}_{2k}(X;Q)$ is split. It suffices to find a cobordism $X \sim X'$ such that $IH^{\bar{m}}_{2k}(X') = 0$, for then X' is the boundary of the cone(X'). By [MH] it is possible to find a symplectic basis for $IH^{\bar{m}}_{2k}(X)$, i.e. a basis in which the matrix of the intersection pairing is

$$\left[\begin{array}{c|c} 0 & I \\ \hline I & 0 \end{array}\right]$$

Siegel shows that every basis element can be represented as an irreducible piecewise linear cycle in X . Let ξ be such a cycle and let $N(\xi)$ be a regular neighborhood of ξ in X , with boundary $\partial N(\xi)$. Siegel now does surgery on X by removing the interior on the neighborhood $N(\xi)$ and replacing it with cone$(N(\xi))$.

Claim. The space $X' = X - N(\xi) \cup \mathrm{cone}(\partial N(\xi))$ is Witt-cobordant to X, and $\dim(IH^{\bar{m}}_{2k}(X')) = \dim(IH^{\bar{m}}_{2k}(X)) - 2$.

Proof of Claim :Notice that X' is homeomorphic to the space obtained from X by collapsing $N(\xi)$ to a point. This kills the homology class $[\xi]$. It turns out that the dual homology class $[\xi]^*$ is also killed : if ξ^* presents a 2k-dimensional cycle in X such that $[\xi].[\xi^*] = 1$, then ξ^* is no longer allowed in $IH^{\bar{m}}_{2k}(X')$ since it would have to pass through the singular point. (A rigorous calculation of $IH_{2k}(X')$ may be made using the formula

$$IH^{\bar{m}}_{2k}(X') = \mathrm{Image}(IH^{\bar{m}}_{2k}(X - N(\xi)) \to IH^{\bar{m}}_{2k}(X,N(\xi))$$

and noticing that this map factors through $IH^{\bar{m}}_{2k}(X))$.

To see that X' is cobordant to X, collapse $N(\xi)$ slowly in $X \times [0,1]$ and add a collar $X' \times [1,1+\varepsilon]$ to the end. Thus we have "filled in" the surgery by adding the cone over $(N(\xi)) \cup \mathrm{cone}\,\partial N(\xi))$ and we must check that the resulting space Y is a Witt space.

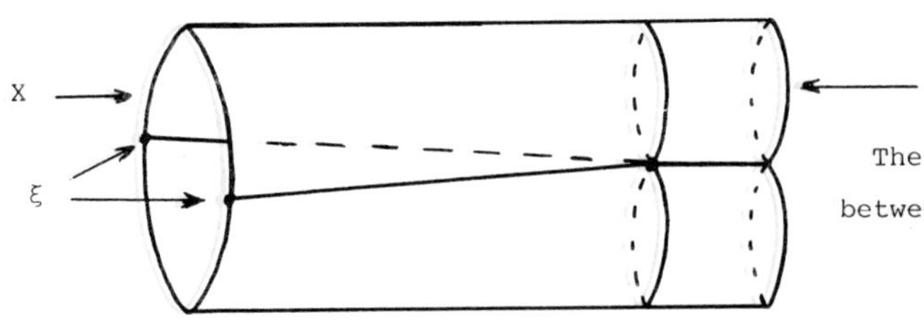

The Cobordism Y , between X and X'

The new singularity of odd codimension in Y is a point whose link is $L = N(\xi)) \cup \mathrm{cone}(N(\xi))$. We must check that $IH^{\bar{m}}_{2k}(L;Q) = 0$. This group is at most one-dimensional and is generated by $[\xi]$: any 2k-dimensional cycle in L must miss the cone point and can therefore be deformed into $N(\xi)$ and then further deformed onto ξ, which was chosen to be irreducible. On the other hand, 0 is a Witt space so $IH^{\bar{m}}_{2k}(L;Q) \times IH^{\bar{m}}_{2k}(L;Q) \to Q$ is nondegenerate. But the original basis was symplectic, so $\xi \cdot \xi = 0$. This means that $IH^{\bar{m}}_{2k}(L;Q) = 0$ as desired.

7. *Remarks.* A similar argument applies to $4k + 2$-dimensional spaces, but in this case a symplectic basis always exists since the intersection pairing is antisymmetric.

A similar theory can be constructed by replacing the coefficients Q with Z/(2) in the definition (*) of Witt spaces. The resulting cobordism groups are 0 in odd dimensions and coincide with $W(Z/(2)) = Z/(2)$ in even dimensions. The invariant is given by the middle intersection homology Euler characteristic, mod 2 .

8. *Problems.*

1. The class $w(X) \in W(Q)$ is a "genus" which gives rise to a characteristic class, part of which is the L class of Hirzebruch and Thom. It would be interesting to study the torsion contributions to this characteristic class, perhaps calculating them for toric varieties.

2. Find a classifying space for this cobordism theory. There does not appear to be any natural notion of a normal microbundle or block bundle of a Witt space.

3. Find an analogous theory using Z coefficients instead of Q coefficients. There is a natural class of pseudomanifolds whose intersection homology satisfies Poincaré duality over the integers [GS]. These are stratified spaces such that

a) $IH^{\bar{m}}(L;Z) = 0$ whenever L is the link of a stratum of odd codimension, $c = \ell + 1$

b) $IH^{\bar{m}}_{\frac{\ell-1}{2}}(L;Z)$ is torsion free whenever L is the link of a stratum of even codimension $c = \ell + 1$.

9. REFERENCES

[GM1] M. GORESKY and R. MACPHERSON, Intersection homology theory. Topology 19 (1980) 135-162

[GM2] M. GORESKY and R. MACPHERSON, Intersection homology II, Inventiones Mathematicae 72 (1983) 77-130

[GS] M. GORESKY and P. SIEGEL, Linking pairings of singular spaces, Commentarii Mathematici Helvetici 58 (1983) 96-110

[MH] J. MILNOR and D. HUSEMOLLER, Symmetric Bilinear Forms, Springer-Verlag, New York 1973

[S] P. SIEGEL, Witt spaces: a geometric cycle theory for KO homology at odd primes, to appear in Am. J. of Mathematics

Intersection Cohomology
Progress in Mathematics, Vol. 50
© 1984 Birkhäuser Boston, Inc.

VIII. LEFSCHETZ FIXED POINT THEOREM AND INTERSECTION HOMOLOGY

by

Mark Goresky and Robert MacPherson

This article is a summary of the essential ingredients in [3]. We will consider a placid self-map with isolated fixed points on a subanalytic pseudomanifold and show that the trace of the induced homomorphism on intersection homology may be interpreted as a sum of certain linking numbers at the fixed points.

In this note all spaces will be compact oriented sub-analytic pseudomanifolds which admit stratifications by even codimension strata. All maps will be subanalytic, and intersection homology will be taken with middle perversity and field coefficients, k.

Definition: A map $f: X \to Y$ is placid if there exists a stratification of Y such that for each stratum S, $\operatorname{codim}_X f^{-1}(S) \geq \operatorname{codim}_Y(S)$.

A placid map induces homomorphisms both ways on intersection homology. Thus we can define the Lefschetz number of a placid self map $f: X \to X$ by

$$IL(f) = \Sigma(-1)^i \operatorname{Tr}(f_*: IH_i(X) \to IH_i(X)) .$$

The Lefschetz fixed point theorem of Verdier [4] implies that $IL(f)$ can be written as a sum of local contributions at the fixed points of f. We wish to study these local contributions.

Intersection Homology of a join

Let L_1 and L_2 be pseudomanifolds of dimension ℓ. Consider the join $L_1 * L_2$. It has a covering by two open sets $\mathrm{cone}(L_1) \times L_2$ and $L_1 \times \mathrm{cone}(L_2)$ whose intersection is $L_1 \times L_2 \times (0,1)$. Using Mayer Vietoris and the Kunneth formula it is easy to see that $IH_\ell(L_1 * L_2) = IH_{\ell+1}(L_1 * L_2) = 0$.

Intersection Homology of $X \times X$

Choose two stratifications of X and stratify $X \times X$ by the product stratification. Let $\bar{p}$ be the (stratum dependent) perverisity which attaches the following number to a stratum $A \times B$ of $X \times X$:

$$p(A \times B) = \begin{cases} \dfrac{\mathrm{cod}(A) + \mathrm{cod}(B) - 1}{2} & \text{if } \mathrm{cod}(A) \neq \mathrm{cod}(B) \\ \mathrm{cod}(A) & \text{if } \mathrm{cod}(A) = \mathrm{cod}(B) \end{cases}$$

Proposition: The natural homomorphism $IH_*^{\bar{m}}(X \times X) \to IH_*^{\bar{p}}(X \times X)$ is an isomorphism.

Proof: In fact the complexes $IC_*^{\bar{m}}(X \times X)$ and $IC_*^{\bar{p}}(X \times X)$ are quasi-isomorphic, as can be seen from Deligne's construction of the complex of sheaves IC (see [2]). If $(x_1,x_2) \in A \times B$ is a point in a stratum where $\mathrm{codim}(A) = \mathrm{codim}(B)$, then the stalk cohomology at (x_1,x_2) of the two complexes $IC^{\bar{m}}$ and $IC^{\bar{p}}$ differ only in one dimension, and in this dimension the offending group is $IH_\ell(L_1 * L_2) = 0$, where L_1 is the link of A and L_2 is the link of B and $\ell = \mathrm{codim}(A) - 1 = \dim(L_1) = \dim(L_2)$. (Note that $L_1 * L_2$ is the link in $X \times X$ of the stratum $A \times B$. We have used the "obstruction sequence" technique from [2] §5.5).

Intersecting the Graph and Diagonal

Let $f: X^n \to X^n$ be a placid map. An easy dimension count shows that the graph

$$G(f) = \{(x, f(x)) \mid x \in X\}$$

is a $(\bar{p}, n)$ - allowable cycle in $IH_n^{\bar{p}}(X \times X)$ which inherits an orientation from that of X because projection, to the first factor $G(f) \to X$ is a homeomorphism. By the preceding proposition, this cycle has a canonical lift to $IH_n^{\bar{m}}(X \times X)$. A similar remark applies to the diagonal, Δ. Thus we can define the intersection product of these two homology classes $[G(f)] . [\Delta]$ to be the image of the class $[G(f)] \otimes [\Delta]$ under the multiplication homomorphism $IH_n(X \times X) \otimes IH_n(X \times X) \to k$ where k is the coefficient field.

Theorem 1. The Lefschetz number $IL(f)$ is equal to the intersection product $[G(f)] . [\Delta]$.

Proof. The proof is the same as in Lefschetz. Choose a basis $\{e_1, \cdots, e_r\}$ for $IH^{\bar{m}}(X)$, and let $\{e_1^*, \cdots, e_r^*\}$ denote the dual basis. Then

$$[\Delta] = \sum_{i=1}^{r} e_i \otimes e_i^*$$

$$[G(f)] = \sum_i \sum_j (-1)^{|e_i|(n-|e_i|)} f_{ij} \, e_i^* \otimes e_j$$

where $|e_i|$ denotes the dimension of e_i and where (f_{ij}) is the matrix of f_* with respect to the basis $\{e_1, \cdots, e_r\}$. Multiplying these two classes gives the alternating sum of traces. (Here we use the sign conventions of Dold [1] § VIII.13.)

Isolated Fixed Points. Suppose $x \in X$ is an isolated fixed point of a placid self map $f: X \to X$. We wish to study the intersection number of the graph of f with the diagonal at the point (x,x). Let L denote the full link of x in X, i.e. the boundary of a distinguished neighborhood of x. Then the full link of (x,x) in $X \times X$ is the join $L*L$. The local intersection number of $G(f)$ with the diagonal Δ may be interpreted as a linking number (in $L*L$) of the cycles

$$G_L(f) = G(f) \cap L*L$$

and

$$\Delta_L(f) = \Delta \cap L*L$$

Let $N(\Delta_L)$ be a regular neighborhood of Δ_L in $L*L$ (which admits a stratum preserving deformation retraction $N(\Delta_L) \to \Delta_L$) and let $[\Delta_L]$ denote the homology class in $IH^{\bar{m}}_{n-1}(N(\Delta_L))$ which is carried by Δ_L. There is a unique homology class $K \in IH_n(L*L, N(\Delta_L))$ such that $\partial_*(K) = [\Delta_L]$ where

$$\partial_*: IH_n(L*L, N(\Delta_L)) \to IH_{n-1}(N(\Delta_L))$$

is the connecting homomorphism (which is an isomorphism because $IH_n(L*L) = IH_{n-1}(L*L) = 0$).

Definition. The linking number of $G_L(f)$ and Δ_L in $L*L$ is the image of $K \otimes [G_L(f)]$ under the nondegenerate intersection pairing

$$IH_n(L*L,\ N(\Delta_L)) \otimes IH_{n-1}(L*L - N(\Delta_L)) \to k\ .$$

Remark. If X is normal then both of these groups are one-dimensional.

Theorem 2. If all the fixed points of a placid map $f: X \to X$ are isolated, then the Lefschetz number of f is the sum over all the fixed points of the local linking numbers of $G_L(f)$ with Δ_L.

Outline of proof. The diagonal $\Delta \subset X \times X$ is homologous to a cycle Δ' which narrowly misses each of the fixed points and which coincides with the chains K (defined above) near each fixed point. But the intersection of the graph $G(f)$ and Δ' occurs in points of K and the intersection numbers of $G(f)$ with K are precisely the linking numbers of $G_L(f)$ with Δ_L.

<u>Definition</u>. The self map f is contracting near an isolated fixed point $x \in X$ if there is a (closed) conical neighborhood $U = \text{cone}(L)$ of x, which contains no other fixed points, such that $f(u)$ is contained in the interior U^o of U.

By carefully examining the cycles $G_L(f)$ and Δ_L it is possible to show

<u>Theorem 3</u>. If f is contracting near a fixed point $x \in X$ then the local contribution at x to the intersection homology Lefschetz number of f is the trace

$$\Sigma(-1)^i \; \text{Tr}(f^*\colon IH_i(X,X-x) \to IH_i(X,X-x))$$

of the induced homomorphism on the stalk cohomology of the intersection homology sheaf.

References

[1] A. Dold, Lectures on Algebraic Topology, Springer Verlag (New York) 1972.

[2] M. Goresky and R. MacPherson, Intersection homology II, Inv. Math. 71 (1983) pp. 77-129.

[3] M. Goresky and R. MacPherson, The Lefschetz fixed point theorem for intersection homology, to appear.

[4] A. Grothendieck and L. Illusie, Formule de Lefschetz, in S. G. A. 5, Springer Lecture Notes in Mathematics # 589, Springer-Verlag, N.Y. 1977.

Intersection Cohomology
Progress in Mathematics, Vol. 50
© 1984 Birkhäuser Boston, Inc.

IX. PROBLEMS AND BIBLIOGRAPHY ON INTERSECTION HOMOLOGY

by M. Goresky and R. MacPherson

A. Intersection Homology - versions of other functors

Before considering various possible extensions of intersection homology (such as intersection K-theory), we wish to reflect on small resolutions [D], [F]. For a small resolution $\pi: \tilde{X} \to X$ there is a canonical isomorphism $H_*(\tilde{X}) \cong IH_*(X)$. We might expect other "intersection functors" to satisfy a similar identity. A severe limitation on the existence of such functors is therefore provided by the existence of spaces which have two different small resolutions.

We illustrate these ideas by showing that it is not possible to find a natural ring structure on intersection homology or to find Chern classes in intersection homology.

Example 1. Let X be the Schubert variety in the Grassmannian $G_2(\mathbb{C}^4)$ consisting of all complex two-planes $V^2 \subset \mathbb{C}^4$ such that $\dim(V^2 \cap \mathbb{C}^2) \geq 1$. This variety X has a singularity at the point $V^2 = \mathbb{C}^2$, and it has a small resolution $\pi_1: \tilde{X}_1 \to X$ where $\tilde{X}_1$ consists of all $(1,2)$-flags $V^1 \subset V^2 \subset \mathbb{C}^4$ such that $V^1 \subset \mathbb{C}^2$. It has a second small resolution $\pi_2: \tilde{X}_2 \to X_2$ which consists of all $(2,3)$ flags $V^2 \subset V^3 \subset \mathbb{C}^4$ such that $\mathbb{C}^2 \subset V^3$. Although $\tilde{X}_1$ and $\tilde{X}_2$ are homeomorphic, the homology isomorphisms

$$H_*(\tilde{X}_1) \xrightarrow{(\pi_1)_*} IH_*(X) \xleftarrow{(\pi_2)_*} H_*(\tilde{X}_2)$$

do not preserve the ring structures, nor (by a calculation of Verdier) do they preserve the Chern classes.

<u>Example 2</u>. Let X be the Schubert variety

$$X = \{V^2 \in G_2(\mathbb{C}^5) \mid \dim(V^2 \cap \mathbb{C}^3) \geq 1\} .$$

Let $\tilde{X}_1$ be the variety of partial flags $V^1 \subset V^2 \subset \mathbb{C}^5$ such that $V_1 \subset \mathbb{C}^2$. Let $\tilde{X}_2$ be the variety of partial flags $V^2 \subset V^4 \subset \mathbb{C}^5$ such that $\mathbb{C}^3 \subset V^4$. Both $\tilde{X}_1$ and $\tilde{X}_2$ are small resolutions of X but their cohomology rings $H^*(\tilde{X}_1)$ and $H^*(\tilde{X}_2)$ are not even abstractly isomorphic.

<u>Remark</u> ([8]). It is not possible in general to find, for any variety X a new space $\pi: \tilde{X} \to X$ such that the intersection homology sheaf of X is quasi-isomorphic to the pushforward $R\pi_*(\underline{\underline{\mathbb{Z}}}_{\tilde{X}})$ of the constant sheaf on X (i.e. such that

$$IH_*(U) \cong H_*(\pi^{-1}(U))$$

for each open set $U \subset X$).

<u>Problem #1</u>. Is there an "intersection homology version" of cobordism theory or K theory (or homotopy theory)? Such a theory should satisfy Poincaré duality (in some suitable sense) and should agree with the usual cobordism or K theory of a small resolution (when one exists). In particular one must determine whether the cobordism or K groups of any two small resolutions coincide.

An obvious attempt at such a theory would be to define $I\,\Omega_i^{\bar{p}}(X)$ to be cobordism classes of continuous maps $f: M \to X$ (from a smooth i-dimensional manifold M to a stratified space X), which are required to satisfy a perversity condition

$$\dim f^{-1}(S) \leq i - \operatorname{cod}(S) + p(\operatorname{cod}(S))$$

whenever S is a statum of X. However it is unclear how to construct a product

$$I\,\Omega_*^{\bar{p}}(X) \otimes I\,\Omega_*^{\bar{q}}(X) \to I\,\Omega_*^{\bar{p}+\bar{q}}(X)$$

and there is no reason to believe that Poincaré duality will hold for such a theory.

B. Relations with Analysis

Let X be a complex projective algebraic variety. Consider the Kaehler metric on the nonsingular part X° of X which is induced from the Fubini-Study metric on the ambient projective space.

Problem #2: In [15] it is conjectured that the closed L^2 differential p-forms modulo the exact L^2 differential p-forms is a finite dimensional group which is canonically isomorphic to $IH_{2n-p}(X)$. It is also conjectured that every such L^2 cohomology class contains a unique harmonic form, and the $\mathbb{C}$-valued harmonic forms can be split into (p,q)-components thus inducing a pure Hodge structure on $IH_*(X)$.

There should also be relations between $IH_*(X)$ and other differential operators on L^2 (see the discussion in [F]).

Problem 3: Does the index of the $\bar{\partial}$ complex on the L^2 differential forms on X° coincide with the arithmetic genus of every resolution of singularities of X? What is the correct generalization of the $\hat{A}$ genus to singular varieties so that it coincides with the index of the L^2 Dirac operator (see [2])?

These L^2 methods may eventually be used to prove that the decomposition formula ([A], [20]) holds for complex analytic maps.

C. Functoriality of Intersection Homology

Problem #4: Find the most general category of spaces and maps (perhaps with additional data) on which intersection homology is functorial.

Remark: Problem #4 has a trivial solution: intersection homology is a functor on the category whose morphisms from

X to Y are continuous maps $f: X \to Y$ together with a "lift" to a sheaf map $f_* \mathcal{IC}^\cdot(X) \to \mathcal{IC}^\cdot(Y)$. The real problem is to find a functor from a more geometric category to this one.

<u>Discussion</u>: (In this discussion we will treat only the middle perversity.) There are several classes of maps $f: X \to Y$ for which one can naturally associate homomorphisms between $IH_*(X)$ and $IH_*(Y)$. (If we restrict to field coefficients every natural homomorphism $IH_*(X) \to IH_*(Y)$ has an adjoint $IH_*(Y) \to IH_*(X)$ so functoriality will be both covariant and contravariant.)

(a) Placid maps: [6] [E] [F] These are not closed under composition, but there is a well defined placid homotopy category and intersection homology is functorial on this category.

(b) Small maps [D]. There is no obvious way to make these into a category. However if the composition of two small maps f and g happens to be small then $f_* g_* = (fg)_*$.

There are examples of diagrams of placid or small maps such that the induced homomorphisms on homology do not commute. The varieties X, $\tilde{X}_1$ and $\tilde{X}_2$ can be taken as in example 1 above, and $\tilde{\tilde{X}}$ can be taken to be the blow up of X at the singular point. Then there is a commutative diagram

$$\begin{array}{ccc} \tilde{\tilde{X}} & \xrightarrow{f_1} & \tilde{X}_1 \\ {\scriptstyle f_2}\downarrow & & \downarrow{\scriptstyle \pi_1} \\ \tilde{X}_2 & \xrightarrow[\pi_2]{} & X \end{array}$$

where f_1 and f_2 are placid, π_1 and π_2 are small, but the induced square on intersection homology does not commute.

(c) Proper algebraic surjections: Deligne has shown that if $f: X \to Y$ is a proper algebraic surjection and if D is a choice of a relatively ample divisor then the direct sum decomposition ([A], [20], [F]) can be made canonical. This procedure produces a map

$$RF_*: IC^{\cdot}_X \to IC^{\cdot}_Y .$$

(d) Two-way morphisms: A possible replacement for functoriality of intersection homology may be a group $IH_*(f)$ defined for appropriately stratified maps $f: X \to Y$, which maps both ways, i.e.

$$IH_*(X) \leftarrow IH_*(f) \to IH_*(Y) .$$

A candidate for $IH_*(f)$ may be constructed using chains on X which satisfy the allowability conditions in X and whose images satisfy the allowability conditions in Y. However it is not clear how to formulate functoriality of this construction, nor is it clear to what extent $IH_*(f)$ is invariant under restratification of X and Y.

(e) The following question is even more speculative: For an oriented map $f: X \to Y$ ([3]) with equidimensional fibres, is there a functor $f^{\bar{m}}: D^b(Y) \to D^b(X)$ (defined on the constructible derived categories of X and Y) with the property that $f^{\bar{m}}(IC^{\cdot}_Y) = IC^{\cdot}_X$? This functor should be "halfway" between the functors $f^* = f^{\bar{0}}$ and $f^! = f^{\bar{t}}$ and the relative orientation should induce natural transformations $f^* \to f^{\bar{m}} \to f^!$. If such a functor exists we might define $IH(f)$ to be the hypercohomology of $f^{\bar{m}}(\underline{\underline{\mathbb{Z}}}_Y)$. In any case one can ask whether there is a bivariant intersection homology theory ([3]).

D. Cobordism Calculations

Fix a (finite dimensional regular Noetherian) coefficient ring R. Let $\bar{p} \le \bar{m}$ be a perversity below the middle one, and let $\bar{q} = \bar{t} - \bar{p}$ be the complementary perversity. We shall say an n-dimensional compact pseudomanifold X satisfies $\bar{p}$ - partial Poincaré duality (over the ring R) provided

(1) the natural sheaf map $IC^{\cdot}_{\bar{p}} \to IC^{\cdot}_{\bar{q}}$ is a quasi-isomorphism, and

(2) the intersection pairing $IC^{\cdot}_{\bar{p}} \times IC^{\cdot}_{\bar{q}} \to \mathcal{D}^{\cdot}[n]$ induces a quasi-isomorphism

$$IC^{\cdot}_{\bar{p}} \to R\,\mathcal{H}om^{\cdot}(IC^{\cdot}_{\bar{q}}, \mathcal{D}^{\cdot})[n] .$$

Problem #5: Compute the cobordism groups $\Omega_*^{\bar{p}}$ of spaces satisfying $\bar{p}$ - partial Poincaré duality.

Discussion: Condition (1) and (2) imply that $IH_*^{\bar{p}}(X)$ satisfies Poincaré duality over R. If R is a field then condition (2) is always satisfied; in general this is a condition on the higher tor groups of the intersection homology groups of links of strata (see [7] for the case $R = \mathbb{Z}$). Condition (1) is equivalent to the following statement: for each stratum of codimension c, if $p(c) < q(c)$ then

$$IH^{\bar{p}}_{p(c)+1}(L;R) = IH^{\bar{p}}_{p(c)+2}(L;R) = \cdots = IH^{\bar{p}}_{q(c)}(L;R) = 0$$

where L is the link of the stratum in question.

Several interesting cases of these cobordism groups are known:

(a) $R = Q$, $\bar{p} = \bar{m}$. These spaces are the Witt spaces of Siegel ([14]) and the cobordism groups are $\Omega_i^{\bar{m}} = 0$ unless $i \equiv 0 \pmod 4$ and $\Omega_{4k}^{\bar{m}} = W(Q)$, the Witt group of the rationals.

(b) $R = \mathbb{Z}/(2)$, $\bar{p} = \bar{m}$: These spaces are "$\mathbb{Z}/(2)$-Witt spaces" and the cobordism groups are $\Omega_{odd} = 0$; $\Omega_{even} = \mathbb{Z}/(2)$.

(c) $R = \mathbb{Z}$, $\bar{p} = \bar{m}$: These spaces are discussed in [7] and the cobordism groups have been shown ([13]) to coincide with the higher Mischenko-Witt groups of the integers.

(d) $R = Q$, $\bar{p} = \bar{0}$: These spaces are rational homology manifolds whose cobordisms groups were computed by Maunder.

(e) $R = Q$, $p(c) = [\frac{c-2}{3}]$. For these spaces $IH_*^{\bar{p}}(X)$ forms a non-associative?) ring because the product of any two classes in $IH_*^{\bar{p}}(X)$ lies in $IH_*^{2\bar{p}}(X) \cong IH_*^{\bar{p}}(X)$.

Is it possible to find explicit constructions for the classifying spaces of these cobordism theories?

There are canonical maps $\Omega_*^{\bar{p}} \to \Omega_*^{\bar{q}}$ whenever $\bar{q} \geq \bar{p}$. One expects these maps to be rationally surjective and to correspond to "killing" various characteristic numbers: as $\bar{p}$ increases it should be harder and harder to define characteristic numbers by intersecting characteristic classes.

There should also be interesting products between the various $\Omega_*^{\bar{p}}$.

Problem #6: Is there a sheaf-theoretic construction of the bordism group $\Omega_i^{\bar{m}}(X)$ of continuous maps from i-dimensional Witt spaces to X? The objects in this theory should be certain equivalence classes of complex of sheaves $\underline{\underline{S}}^{\cdot}$ on X together with a Verdier dual pairing $\underline{\underline{S}}^{\cdot} \otimes \underline{\underline{S}}^{\cdot} \to \underline{\underline{\mathbb{D}}}_X^{\cdot}$. If $f: Y \to X$ is an element of $\Omega_i^{\bar{m}}(X)$ then the corresponding sheaf is $\underline{\underline{S}}^{\cdot} = Rf_* \quad {}^{\cdot}_{Y}$. If X is a point then we should recover the higher Mischenko-Witt groups.

E. Intersection Homology of Real Algebraic Varieties

Problem #7: Is there a self-dual $\mathbb{Z}/(2)$-generalization of intersection homology for real algebraic varieties?

Discussion: It would not be a purely topological invariant-for example it would give different groups for the following two spaces:

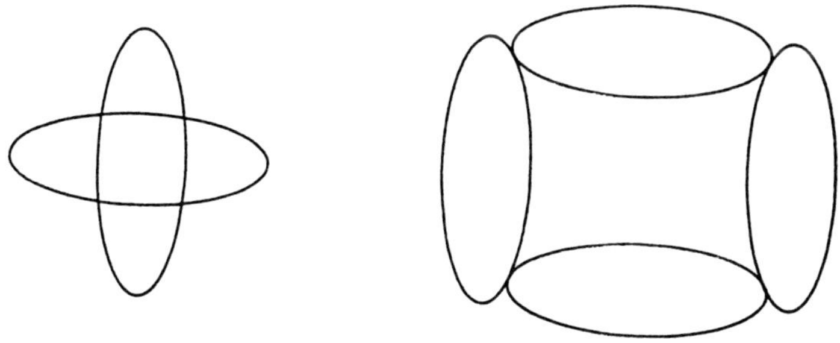

If we denote this proposed group by IJ_* we would expect $IJ_1(X) = \mathbb{Z}/(2) \oplus \mathbb{Z}/(2)$ while $IJ_1(Y) = \mathbb{Z}/(2) \oplus \mathbb{Z}(2) \oplus \mathbb{Z}/(2) \oplus \mathbb{Z}(2)$.

If the real algebraic variety X has a (real) small resolution $\tilde{X}$ then $IJ_*(X)$ should agree with $H_*(\tilde{X})$. The example of a real Schubert variety is suggestive: Let X be the real version of example 1 from section A. Then, $IJ_*(X) = H_*(\tilde{X};\mathbb{Z}/(2))$ which is $\mathbb{Z}/(2)$ in dimensions 0 and 3, and is $\mathbb{Z}/(2) \oplus \mathbb{Z}/(2)$ in dimensions 1 and 2. Notice that this is not $IH_*^{\bar{p}}(X;\mathbb{Z}/(2))$ for any perversity $\bar{p}$. A preliminary problem would be to determine whether $H_*(\tilde{X};\mathbb{Z}/(2))$ is the same for all small resolutions $\tilde{X}$ of a real algebraic variety X.

F. Intersection Homology of Noncompact Varieties

The following question was asked by D. Kazhdan:

Problem #8: Is there a generalization of intersection homology which applies to a noncompact variety and which satisfies Poincaré duality and hard Lefschetz?

We remark that it is easy to find self dual homology groups $F_i(X)$ associated to a non-compact variety. For example

$$F_i(X) = \text{Image } (H_i(X) \to H_i^{BM}(X))$$

where $H_i^{BM}(X)$ denotes the (Borel-Moore) homology with closed supports. If X is nonsingular then this group is dual to $F_{2n-i}(X)$ and Deligne has pointed out that it even has a pure Hodge structure. However, it does not satisfy hard Lefschetz.

G. Orbits of Reductive Group Actions

Intersection homology has been computed for Schubert varieties ([41], [42], [40]), toric varieties ([E]) and for $K_{\mathbb{C}}$ orbits on the variety of Borel subgroups of a complex reductive group [47], and nilpotent varieties ([36], [37]). In each of these cases both the global and the local intersection homology turns out to be zero in odd degrees.

Problem #9: Explain this phenomenon. Is it also true for any space on which a reductive group acts with finitely many orbits?

H. Characteristic Numbers of Algebraic Varieties

Problem #10: Which characteristic numbers can be defined for all algebraic varieties, in such a way that they coincide with the (usual) characteristic numbers of $\tilde{X}$ whenever $\tilde{X} \to X$ is a small resolution?

Discussion: This list is known to contain the Euler characteristic, the signature (or "L-genus"), and the arithmetic (or Todd) genus. Do all algebraic cycles lift (rationally) to intersection homology? If so, and if the lifts are sufficiently canonical, can we then multiply Chern classes to create more characteristic numbers?

I. Miscellaneous

Problem #11: Is there a category of spaces, maps and homotopies, and a "classifying space" B so that $IH_i(X)$ can be interpreted as homotopy classes of maps from X to B?

Problem #12: Find a stratification independent constructive definition of intersection homology. (Note that a stratification independent characterization is given in [D]).

Bibliography for Intersection Homology

General References

[A] A. A. Beilinson, J. Bernstein, P. Deligne, Faisceaux Pervers, Soc. Math. de France, Asteristique #100 (1983).

[B] J. L. Brylinski, (Co)-homologie d'intersection et faisceaux pervers. Sem. Bourbaki #585, Soc. Math. de France Asterisque 92-93 (1982), pp. 129-158.

[C] M. Goresky and R. MacPherson, Intersection homology theory, Topology 19 (1980), pp. 135-162.

[D] M. Goresky and R. MacPherson, Intersection Homology II, Inv. Math. 72 (1983), pp. 77-130.

[E] R. MacPherson, Intersection Homology (Hermann Weyl Lectures) to appear in Annals of Mathematics Studies, Princeton University Press.

[F] R. MacPherson, Global questions in the topology of singular spaces, lecture delivered to ICM (Warsaw, Poland), Aug. 1983.

[G] T. A. Springer, Quelques applications de la cohomologie d' intersection. Séminaire Bourbaki #589, Soc. Math. de France Astérisque # 92-93 (1982) pp. 249-274.

[H] Seminar on Intersection Homology, Bern Switzerland (A. Borel, ed). This seminar.

[I] Analyse et Topologie sur les Espaces Singuliers, Soc. Math. de France Astérisque #101, 102.

Topological Papers

[1] J. Cheeger, On the Hodge theory of Riemannian pseudo-manifolds. Proc. Symposia Pure Math. 36, Providence: AMS (1980), pp. 91-146.

[2] A. Chou, The Dirac operator on spaces with conical singularities and positive scalar curvature, to appear.

[3] W. Fulton and R. MacPherson, Categorical Framework for the study of singular spaces. Mem. Amer. Math. Soc. 243 (1981) A.M.S. Providence RI (1981).

[4] M. Goresky, Intersection homology operations, to appear in Comment. Math. Helv.

[5] M. Goresky and R. MacPherson, La dualité de Poincaré pour les espaces singuliers, C.R. Acad. Sci. t. 284 (Serie A) (1977), 1549-1551.

[6] M. Goresky and R. MacPherson, The Lefschetz fixed point theorem and intersection homology, to appear. (See also [H]).

[7] M. Goresky and P. Siegel, Linking pairings on singular spaces, Comment. Math. Helv. 58 (1983) pp. 96-110.

[8] N. Habegger, Obstructions to immersions in intersection homology theory, preprint. Univ. of Geneva, 1983.

[9] H. King, Intersection homology and homology manifolds. Topology 21 (1982).

[10] H. King, Topological invariance of intersection homology without sheaves. (to appear in Topology).

[11] R. MacPherson and K. Vilonen, Construction elementaire des faisceaux pervers, preprint, 1983.

[12] M. Nagase, L^2 cohomology and intersection cohomology of stratified spaces. Duke Math J. 50 (1983), pp. 329-368.

[13] W. Pardon, Cobordism groups of integral Witt spaces, to appear.

[14] P. Siegel, Witt spaces, a geometric cycle theory for KO homology at odd primes. Ph.D. thesis (M.I.T), 1979. (to appear in Amer. J. of Math.)

Papers on Intersection Homology and Algebraic Geometry

[15] J. Cheeger, M. Goresky, and R. MacPherson, L^2-cohomology and intersection homology for singular algebraic varieties, proceedings of year in differential geometry, I.A.S., S. Yau, ed. (1981) Annals of Math. Studies, Princeton Univ. Press.

[16] P. Deligne, Pureté de la cohomologie de MacPherson-Goresky, d'après un expose de O. Gabber, rédigé par P. Deligne, IHES preprint, Fév. 1981.

[17] P. Deligne, Applications de la Pureté, Proc. of D-modules et singularités, Astérisque, to appear.

[18] W. Fulton and R. Lazarsfeld, The numerical positivity of ample vector bundles. Ann. Math. 118 (1983) pp. 35-60.

[19] M. Goresky and R. MacPherson, Morse theory and intersection homology, in [I] pp. 135-192

[20] M. Goresky and R. MacPherson, On the topology of complex algebraic maps, Algebraic Geometry-Proceedings, La Rabida, Springer Lect. Notes in Math. No. 961 (1982), pp. 119-129.

[21] W. C. Hsiang and V. Pati, L^2 cohomology of normal algebraic surfaces, preprint, Princeton Univ., 1983.

[22] J. Steenbrink, Mixed Hodge structures associated with isolated singularities, in Singularities, Proc. of Symp. in pure Math. vol. 40 #2 pp. 513-536. Amer. Math. Soc. Providence RI (1983).

[23] J.L. Verdier, Specialization de faisceaux et Monodromie Modérée, in [I], pp. 332-364.

Papers on Intersection Homology and D Modules
(see also [11], [34], [38], [39], [40])

[24] J. Bernstein, Algebraic theory of D-modules, proc. of "D-modules et singularities", Astérisque, to appear.

[25] J.-L. Brylinski, Modules holonomes a singularitiés régulières et filtration de Hodge I, Algebraic Geometry (Proceedings La Rabida) Lect. Notes in Math. No. 961, Springer-Verlag (1982) pp. 1-21.

[26] J.L. Brylinski, Modules holonomes à singularités régulières et filtrations de Hodge II, in [I] pp. 75-117.

[27] J.L. Brylinski, Transformations canoniques, dualité projective, transformation de Fourier, et sommes trigonmetriques, to appear in Astérisque.

[28] J.L. Brylinski, A. Dubson, and M. Kashiwara, Formule de l'indice pour les modules holnomes et obstruction d'Euler locale, C.R.A.S., (1981)

[29] J.L. Brylinski, B. Malgrange, and J.-L. Verdier, Transformations de Fourier géométrique I, C.R. Acad. Sci. Paris 297 (1983), 55-58.

[30] A. Galligo, M. Granger, and P. Maisonobe, D-modules et faisceaux pervers dont le support singulier est un croisement normal, preprint, Univ. of Nice (1983).

[31] M. Kashiwara, The Riemann-Hilbert problem, preprint, R.I.M.S. (1983).

[32] Le D.T. and Z. Mebkhout, Introduction to linear differential systems, in Singularities Proc. of Symp. in pure Math. #40 (Part 2) pp. 31-64, Amer. Math. Soc. Providence, RI (1983).

[33] T. Oda, Introduction to algebraic analysis on complex manifolds, Adv. Stud. in Pure Math. 1 (1983) pp. 29-48, Japan Math. Soc.

Papers on Intersection Homology and Group Representation Theory

[34] A. Beilinson and J. Bernstein, Localisation des G-modules, C.R.A.S., No. 292 (1981), pp. 15-18.

[35] A. Borel, L^2 cohomology and intersection cohomology of certain arithmetic varieties, Proc. of E. Noether Symposium at Bryn Mawr, Springer Verlag.

[36] W. Borho and R. MacPherson, Representations des groupes de Weyl et homologie d'intersection pour les varieties de nilpotents, C.R. Acad. Sc. Paris, 292 (1981), pp. 707-710.

[37] W. Borho and R. MacPherson, Partial resolution of nilpotent varieties, in [I] pp. 23-74.

[38] J. L. Brylinski and M. Kashiwara, Démonstration de la Conjecture de Kazhdan et Lusztig sur les modules de Verma (C.R. Acad. Sci. Paris t.291 (1980) pp. 373-376).

[39] J. L. Brylinski and M. Kashiwara, Kazhdan-Lusztig conjecture and holonomic systems, Invent. math. 64 (1981), pp. 387-410.

[40] S. Gelfand, R. MacPherson, Verma modules and Schubert cells; a dictionary, Séminaire d'Algèbre. Lecture Notes in Math., vol 924, Berlin-Heidelberg-New York: Springer 1982, pp. 1-50.

[41] D. Kazhdan and G. Lusztig, Representations of Coxeter groups and Hecke algebras, Inv. Math. 53 (1979), pp. 165-184.

[42] D. Kazhdan and G. Lusztig, Schubert varieties and Poincaré duality, Proc. Symp. Pure Math. vol. 36, pp. 185-203, Amer. Math. Soc. (1980).

[43] G. Lusztig, Representations of Hecke algebras and Coxeter groups (preprint MIT 1982)-edited by D. King.

[44] G. Lusztig, Green polynomials and singularities of unipotent classes, Adv. in Math. 42 (1981), pp. 169-178.

[45] G. Lusztig, Some problems in the representation theory of finite Chevalley groups, Proc. Symp. Pure Math. vol. 37, Amer. Math. Soc. (1980), pp. 313-317.

[46] G. Lusztig, Equivariant sheaves on reductive groups I, preprint, M.I.T. 1982.

[47] G. Lusztig and D. Vogan, Singularities of closures of K-orbits on flag manifolds, Inv. Math. 71, No. 2 (1983), pp. 365-380.

[48] G. Lusztig, Intersection homology complexes on reductive groups, Inv. Math. 75 (1984), 205-272.

[49] T.A. Springer, Trigonometric sums, Green functions of finite groups and representations of Weyl groups, Inv. Math. 36 (1976), pp. 173-207.

[50] A. Zelevinsky, A p-adic analogue of the Kazhdan-Lusztig conjecture (in Russian), Funct. Anal. and its Appl. 15 (1981), pp. 9-21.

[51] S. Zucker, Hodge theory with degenerating coefficients I, Ann. of Math. 109 (1979), pp. 415-476.

[52] S. Zucker, L_2-cohomology of warped products and arithmetic groups. Inventiones Math. 70 (1982), pp. 169-218.

[53] S. Zucker, L_2-cohomology and intersection homology of locally symmetric varieties, Proc. Symp. pure math. #40 A.M.S. (1983).

[54] S. Zucker, L_2-cohomology and intersection homology of locally summetric varieties II. preprint (1983).

[55] S. Zucker, Hodge theory and arithmetic groups, in [I] pp. 365-381.

AUTHORS' ADDRESSES

Norbert A'Campo
Mathematisches Institut
Universität Basel
8051 Basel, Switzerland

Armand Borel
Mathematisches Forschungsinstitut
ETH-Zentrum
8092 Zürich, Switzerland

The Institute for advanced study
School of Mathematics
Princeton, NJ 08540, USA

Mark Goresky
Northeastern University
Department of Mathematics
Boston, MA 02115, USA

Pierre-P. Grivel
Université de Genève
Faculté des Sciences
Section de Mathématiques
1211 Genève 24, Switzerland

André Haefliger
Université de Genève
Faculté des Sciences
Section de Mathématiques
1211 Genève 24, Switzerland

Nathan Habegger
Yale University
Department of Mathematics
New Haven, CO 06520, USA

Robert MacPherson
Brown University
Department of Mathematics
Providence, R.I. 02912, USA

Nicolas Spaltenstein
Mathematisches Forschungsinstitut
ETH-Zentrum
8092 Zürich, Switzerland

Progress in Mathematics
Edited by J. Coates and S. Helgason

Progress in Physics
Edited by A. Jaffe and D. Ruelle

- A collection of research-oriented monographs, reports, notes arising from lectures or seminars,
- Quickly published concurrent with research,
- Easily accessible through international distribution facilities,
- Reasonably priced,
- Reporting research developments combining original results with an expository treatment of the particular subject area,
- A contribution to the international scientific community: for colleagues and for graduate students who are seeking current information and directions in their graduate and post-graduate work.

Manuscripts

Manuscripts should be no less than 100 and preferably no more than 500 pages in length.

They are reproduced by a photographic process and therefore must be typed with extreme care. Symbols not on the typewriter should be inserted by hand in indelible black ink. Corrections to the typescript should be made by pasting in the new text or painting out errors with white correction fluid.

The typescript is reduced slightly (75%) in size during reproduction; best results will not be obtained unless the text on any one page is kept within the overall limit of 6 × 9½ in (16 × 24 cm). On request, the publisher will supply special paper with the typing area outlined.

Manuscripts should be sent to the editors or directly to:
Birkhäuser Boston, Inc., P.O. Box 2007,
Cambridge, MA 02139 (USA)

PROGRESS IN MATHEMATICS
Already published

PM 1
Quadratic Forms in Infinite-Dimensional Vector Spaces
Herbert Gross
ISBN 3-7643-1111-8
432 pages, paperback

PM 2
Singularités des systèmes différentiels de Gauss-Manin
Frédéric Pham
ISBN 3-7643-3002-3
346 pages, paperback

PM 3
Vector bundles on Complex Projective Spaces
C. Okonek, M. Schneider, H. Spindler
ISBN 3-7643-3000-7
396 pages, paperback

PM 4
Complex Approximation, Proceedings, Quebec, Canada, July 3–8, 1978
Edited by Bernard Aupetit
ISBN 3-7643-3004-X
128 pages, paperback

PM 5
The Radon Transform
Sigurdur Helgason
ISBN 3-7643-3006-6
207 pages, hardcover

PM 6
The Weil Representation, Maslov Index and Theta Series
Gérard Lion, Michèle Vergne
ISBN 3-7643-3007-4
348 pages, paperback

PM 7
Vector Bundles and Differential Equations Proceedings, Nice, France, June 12–17, 1979
Edited by André Hirschowitz
ISBN 3-7643-3022-8
256 pages, paperback

PM 8
Dynamical Systems, C.I.M.E. Lectures, Bressanone, Italy, June 1978
John Guckenheimer, Jürgen Moser, Sheldon E. Newhouse
ISBN 3-7643-3024-4
305 pages, hardcover

PM 9
Linear Algebraic Groups
T. A. Springer
ISBN 3-7643-3029-5
314 pages, hardcover

PM 10
Ergodic Theory and Dynamical Systems I
A Katok
ISBN 3-7643-3036-8
346 pages, hardcover

PM 11
18th Scandinavian Congress of Mathematicians, Aarhus, Denmark, 1980
Edited by Erik Balslev
ISBN 3-7643-3040-6
526 pages, hardcover

PM 12
Séminaire de Théorie des Nombres, Paris 1979–80
Edited by Marie-José Bertin
ISBN 3-7643-3035-X
404 pages, hardcover

PM 13
Topics in Harmonic Analysis on Homogeneous Spaces
Sigurdur Helgason
ISBN 3-7643-3051-1
152 pages, hardcover

PM 14
Manifolds and Lie Groups, Papers in Honor of Yozô Matsushima
Edited by J. Hano, A. Marimoto, S. Murakami, K. Okamoto, and H. Ozeki
ISBN 3-7643-3053-8
476 pages, hardcover

PM 15
Representations of Real Reductive Lie Groups
David A. Vogan, Jr.
ISBN 3-7643-3037-6
776 pages, hardcover

PM 16
Rational Homotopy Theory and Differential Forms
Phillip A. Griffiths, John W. Morgan
ISBN 3-7643-3041-4
258 pages, hardcover

PM 17
Triangular Products of Group Representations and their Applications
S.M. Vovsi
ISBN 3-7643-3062-7
142 pages, hardcover

PM 18
Géométrie Analytique Rigide et Applications
Jean Fresnel, Marius van der Put
ISBN 3-7643-3069-4
232 pages, hardcover

PM 19
Periods of Hilbert Modular Surfaces
Takayuki Oda
ISBN 3-7643-3084-8
144 pages, hardcover

PM 20
Arithmetic on Modular Curves
Glenn Stevens
ISBN 3-7643-3088-0
236 pages, hardcover

PM 21
Ergodic Theory and Dynamical Systems II
A. Katok, editor
ISBN 3-7643-3096-1
226 pages, hardcover

PM 22
Séminaire de Théorie des Nombres, Paris 1980–81
Marie-José Bertin, editor
ISBN 3-7643-3066-X
374 pages, hardcover

PM 23
Adeles and Algebraic Groups
A. Weil
ISBN 3-7643-3092-9
138 pages, hardcover

PM 24
Enumerative Geometry and Classical Algebraic Geometry
Patrick Le Barz, Yves Hervier, editors
ISBN 3-7643-3106-2
260 pages, hardcover

PM 25
Exterior Differential Systems and the Calculus of Variations
Phillip A. Griffiths
ISBN 3-7643-3103-8
349 pages, hardcover

PM 26
Number Theory Related to Fermat's Last Theorem
Neal Koblitz, editor
ISBN 3-7643-3104-6
376 pages, hardcover

PM 27
Differential Geometric Control Theory
Roger W. Brockett, Richard S. Millman, Hector J. Sussmann, editors
ISBN 3-7643-3091-0
349 pages, hardcover

PM 28
Tata Lectures on Theta I
David Mumford
ISBN 3-7643-3109-7
254 pages, hardcover

PM 29
Birational Geometry of Degenerations
Robert Friedman and David R. Morrison, editors
ISBN 3-7643-3111-9
410 pages, hardcover

PM 30
CR Submanifolds of Kaehlerian and Sasakian Manifolds
Kentaro Yano, Masahiro Kon
ISBN 3-7643-3119-4
223 pages, hardcover

PM 31
Approximations Diophantiennes et Nombres Transcendants
D. Bertrand and M. Waldschmidt, editors
ISBN 3-7643-3120-8
349 pages, hardcover

PM 32
Differential Geometry
Robert Brooks, Alfred Gray, Bruce L. Reinhart, editors
ISBN 3-7643-3134-8
267 pages, hardcover

PM 33
Uniqueness and Non-Uniqueness in the Cauchy Problem
Claude Zuily
ISBN 3-7643-3121-6,
185 pages, hardcover

PM 34
Systems of Microdifferential Equations
Masaki Kashiwara
ISBN 0-8176-3138-0
ISBN 3-7643-3138-0
182 pages, hardcover

PM 35
Arithmetic and Geometry. Papers Dedicated to I. R. Shafarevich on the Occasion of His Sixtieth Birthday. Volume I: Arithmetic
Michael Artin, John Tate, editors
ISBN 3-7643-3132-1
373 pages, hardcover

PM 36
Arithmetic and Geometry. Papers Dedicated to I. R. Shafarevich on the Occasion of His Sixtieth Birthday. Volume II: Geometry
Michael Artin, John Tate, editors
ISBN 3-7643-3133-X
495 pages, hardcover

PM 37
Mathématique et Physique
Louis Boutet de Monvel, Adrien Douady, Jean-Louis Verdier, editors
ISBN 0-8176-3154-2
ISBN 3-7643-3154-2
454 pages, hardcover

PM 38
Séminaire de Théorie des Nombres, Paris 1981–82
Marie-José Bertin, editor
ISBN 0-8176-3155-0
ISBN 3-7643-3155-0
359 pages, hardcover

PM 39
Classical Algebraic and Analytic Manifolds
Kenji Ueno, editor
ISBN 0-8176-3137-2
ISBN 3-7643-3137-2
644 pages, hardcover

PM 40
Representation Theory of Reductive Groups
P. C. Trombi, editor
ISBN 0-8176-3135-6
ISBN 3-7643-3135-6
308 pages, hardcover

PM 41
Combinatorics and Commutative Algebra
Richard P. Stanley
ISBN 0-8176-3112-7
ISBN 3-7643-3112-7
102 pages, hardcover

PM 42
Théorèmes de Bertini et Applications
Jean-Pierre Jouanolou
ISBN 0-8176-3164-X
ISBN 3-7643-3164-X
140 pages, hardcover

PM 43
Tata Lectures on Theta II
David Mumford
ISBN 0-8176-3110-0
ISBN 3-7643-3110-0
293 pages, hardcover

PM 44
Infinite Dimensional Lie Algebras
Victor G. Kac
ISBN 0-8176-3118-6
ISBN 3-7643-3118-6
268 pages, hardcover

PM 45
Large Deviations and the Malliavin Calculus
Jean-Michel Bismut
ISBN 0-8176-3220-4
ISBN 3-7643-3220-4
230 pages, hardcover

PM 46
Automorphic Forms of Several Variables
Taniguchi Symposium, Katata, 1983
Ichiro Satake and Yasuo Morita, editors
ISBN 0-8176-3172-0
ISBN 3-7643-3172-0
399 pages, hardcover

PM 47
Les Conjectures de Stark sur les Fonctions L d'Artin en s = 0
John Tate
ISBN 0-8176-3188-7
ISBN 3-7643-3188-7
153 pages, hardcover

PM 48
Classgroups and Hermitian Modules
A. Fröhlich
ISBN 0-8176-3182-8
ISBN 3-7643-3182-8
250 pages, hardcover

PM 49
Hyperfunctions and Harmonic Analysis on Symmetric Spaces
Henrik Schlichtkrull
ISBN 0-8176-3215-8
ISBN 3-7643-3215-8
205 pages, hardcover

UNIVERSITY LIBRARY
DURHAM

3 0104 00243190 0

ACCESSION No. AMM 222

This book should be returned on or before the last date shown below, unless it is previously recalled by the Library when it must be returned immediately.

Fines are incurred on overdue books (see Library Regulations).

28 MAY 19

30/8/

23 FEB 19

27 FEB 19

28 MAY